The Sorting Society

The Sorting Society

The Ethics of Genetic Screening and Therapy

Edited by

Loane Skene

and

Janna Thompson

CAMBRIDGE
UNIVERSITY PRESS

CAMBRIDGE UNIVERSITY PRESS
Cambridge, New York, Melbourne, Madrid, Cape Town, Singapore, São Paulo, Delhi

Cambridge University Press
The Edinburgh Building, Cambridge CB2 8RU, UK

Published in the United States of America by Cambridge University Press, New York

www.cambridge.org
Information on this title: www.cambridge.org/9780521689847

First published 2008

Printed in the United Kingdom at the University Press, Cambridge

A catalogue record for this publication is available from the British Library

Library of Congress Cataloguing in Publication data
The sorting society : the ethics of genetic screening and therapy / [edited by]
    Loane Skene, Janna Thompson.
      p. ; cm.
    Includes bibliographical references and index.
    ISBN 978-0-521-68984-7 (pbk.)
    1. Genetic screening–Moral and ethical aspects.   2. Gene therapy–Moral and ethical aspects.
    I. Skene, Loane.   II. Thompson, Janna, 1942–   [DNLM: 1. Genetic Engineering–ethics.
    2. Cloning, Organism–ethics.   3. Genetic Screening–ethics.   4. Reproductive Techniques,
    Assisted–ethics. WB 60 S714 2008]
    RB155.65.S67 2008
    362.196′04207–dc22   2008014091

ISBN 978-0-521-68984-7 paperback

Contents

Contributors

Agnes Bankier Director and Professor, Victorian Clinical Genetics Services, Murdoch Children's Research Institute, Victoria, Australia.

Leslie Cannold Honorary Senior Lecturer, Centre for Gender and Medicine, Monash University, Victoria, Australia.

David Cram Director of Molecular Genetic Services, Monash IVF, Monash Surgical Private Hospital, Clayton, Melbourne, Australia.

Edgar Dahl Senior Research Fellow, Centre for Obstetrics and Gynaecology, University of Giessen, Germany.

Lynn Gillam Centre for Health and Society and Centre for Applied Philosophy and Public Ethics, University of Melbourne; and Murdoch Children's Research Institute, Victoria, Australia.

Mianna Lotz Department of Philosophy, Macquarie University, Sydney, Australia.

David Neil School of English Literatures, Philosophy and Languages, University of Wollongong, New South Wales, Australia.

Julian Savulescu Uehiro Chair in Practical Ethics, Director, Oxford Uehiro Centre for Practical Ethics, University of Oxford, Oxford, UK.

Loane Skene Professor of Law, University of Melbourne, Victoria, Australia.

Rob Sparrow Centre for Human Bioethics, Monash University, Victoria, Australia.

Janna Thompson Department of Philosophy, La Trobe University, Melbourne, Australia.

Preface

The contents and title of this book were inspired by a workshop held at Melbourne University in November 2005 under the auspices of the Australian Research Council Centre for Applied Philosophy and Public Ethics. The workshop brought together experts from law, medicine and philosophy. Many of those who participated are the authors of chapters in this collection.

The 'sorting society' expresses what many people believe will be the outcome of advances in genetic technology: a society in which gender and many characteristics of children are no longer the result of genetic luck but of deliberate selection. The book focuses on the ethical, legal and social issues raised by this technology. Is the prospect of a sorting society something that we should welcome or deplore? Do concerns about how parents or societies might exercise the choice given to them by genetic technology give us reason to restrain its creation or use, and if so how? Would a sorting society increase the freedom of parents and the wellbeing of children or would it undermine values that are central to a liberal democratic society? Would it adversely affect relationships between parents and children or the prospects for future generations?

These are questions of the most profound significance, bearing on the world in which our children and their children and grandchildren will live. Citizens as well as experts need to engage in wise reflection about the development of, use of and restrictions on genetic technology. This book is meant to be a contribution and stimulus to a debate which is likely to become more and more urgent in coming years.

The contributors are not in one mind about the prospects for genetic technology or its ethical implications. The book brings together a range of positions and considerations. Like the workshop from which it originated it is interdisciplinary. Most of the authors are philosophers but experts on genetics and law are also contributors. Our aim as editors is to present the issues in a form readily accessible to readers with no prior knowledge of genetic testing and its uses. We hope that this book will be of interest to philosophers, political commentators, scientists, lawyers, people with genetic conditions and their families – indeed to anyone concerned to be well informed about one of the major issues of our time.

Introduction

Loane Skene and Janna Thompson

Techniques are now available to screen fetuses for serious genetic disorders and, in the future, more and better means will be available to determine their susceptibility to disorders of lesser kinds, including those which occur later in life. It is now technically possible for parents to choose the sex of their child. In the future they may be able to choose other genetically carried or influenced characteristics of their offspring: height, body shape, and perhaps even such things as musical talent, intelligence and emotional traits. It is now technically possible to clone sheep and other animals. In the future people may have the option of cloning their offspring. All of these techniques, both existing and imagined in the future, raise concerns about the development of a 'sorting society', in which parents are able to choose the children they will – and will not – have. This possibility raises serious ethical, medical and legal issues which are discussed in turn by the authors of this book.

There is widespread support for prenatal tests which give prospective parents the opportunity to find out if their child will have a serious genetic disability, thus giving them the choice of terminating a pregnancy if the fetus is defective. Nevertheless, genetic screening has its critics. Some of these are opponents of abortion who think that even genetically defective fetuses have a right to life. Others worry more about the social implications of screening. Does a practice of detecting and eliminating 'defective' fetuses encourage a negative attitude towards people with disabilities? Is it likely to reduce the willingness of the public to provide support for disabled people and for parents who choose to have a severely disabled child? Is it likely to lead to a reduction in research to improve life prospects of severely disabled people?

However, most ethical debate centres on issues raised by existing and future possibilities for genetic manipulation. Preconception sex selection is widely criticized, especially by feminists, and some governments have banned

The Sorting Society: The Ethics of Genetic Screening and Therapy, ed. Loane Skene and Janna Thompson. Published by Cambridge University Press. © Cambridge University Press 2008.

it. The prospect of cloned or 'designer babies' – offspring with a genotype selected by parents – raises difficult questions about the responsibility of parents and medical practitioners, the welfare of children and the future direction of society. Some authors in this volume focus on these issues.

One of the most important ethical considerations concerns medical risk. Cloning is an imperfect and risky technique. Perfecting this technique in the case of animals would not necessarily make it less risky to clone a human being. A cloned child might be subject to serious disorders. It might suffer from premature aging, like Dolly, the first cloned sheep; or from disorders that cannot currently be foreseen or predicted, as is the case for other prospective technologies of genetic manipulation. Even if these technologies can be perfected so that they are relatively risk free, the fact remains that producing the first generations of clones or designer babies will be experiments that could go disastrously wrong. The risks have to be weighed against whatever benefits genetic manipulation is supposed to produce. In her chapter Lynn Gillam explores the advantages that might be derived from the use of cloning to avoid having children with disabilities, but concludes that the benefits do not justify the risks.

Let us assume that in the future the risks will somehow be eliminated and that technologies will be available that would enable parents not only to produce children free of minor as well as major genetic defects but also to select for desired physical, mental and emotional characteristics in their offspring. Would the use of such technologies be ethically acceptable? There are two strong reasons for answering 'yes'.

The first is the assumption that parents should be free to make decisions about having children and what kind of children to have. Genetic screening is widely accepted because it respects the exercise of reproductive freedom. As Agnes Bankier explains, an effort is made by medical practitioners to ensure that those screened have given their informed consent. What parents do about a detected abnormality in their fetus is also up to them, subject to the law. The idea that parents have a 'right' to reproductive freedom can be extended to more controversial cases. Edgar Dahl argues that this freedom should be available to people who want to choose the sex of their children. The same right could be claimed by those who want to choose other characteristics for their offspring.

The second argument for the use of genetic technology focuses on the wellbeing of children. A child with a serious genetic defect is likely to lead a short and miserable life. This seems a good enough moral reason to avoid bringing such a child into the world. In general, it is surely morally better to have healthy children with mental and physical characteristics that will make their lives go well, rather than children who have genetic disadvantages. If this is accepted then it seems that we can hardly avoid an endorsement of genetic manipulation that is much stronger in its implications than that provided by an appeal to reproductive freedom. If it is morally better to bring a healthy rather than disabled child into the world, then parents who could make use of genetic technology to avoid having a disabled child but choose not to use it are doing something that is morally worse. Julian Savulescu provocatively argues that parents have a moral duty to have the best children that they can: children who are healthy, happy, intelligent, emotionally stable and able to make use of the opportunities that their society makes available. Good parents strive to bring about this result by upbringing and education. According to his reasoning, once a risk-free technology of genetic manipulation becomes available, it will be morally imperative for parents to use it.

This conclusion is likely to be resisted on a number of grounds. For one thing it is in tension with the idea that parents should have reproductive freedom. Those who think that something is a moral imperative are not necessarily asking for the state to enforce it. There may be good reasons for not doing so. But the state already interferes with parental prerogatives for the sake of the wellbeing of children; for example, it will not assist parents who want to use genetic screening to ensure that they have a deaf child or one with achondroplasia (dwarfism), to resemble the parents. It is not unreasonable to suppose that in an age of genetic technology the state will become increasingly involved in determining what kind of children parents are 'allowed' to have. The possibility that the state might use the technology to create a world like that described in Aldous Huxley's novel, *Brave New World*[1] (or worse) is a nightmare invoked by some critics. But even if there is little reason to think that that would happen, the costs of making the use of the technology imperative, if only to avoid genetically caused disabilities, might be considerable.

Leslie Cannold points out that reproductive therapy is intrusive and time consuming. One major reason for insisting on reproductive freedom is the role of women in reproduction. It is their bodies that are subject to risk and manipulation. It is they who have to suffer the intrusions that are imposed by reproductive technology. Even if the moral imperative to have the best possible children never becomes a legal requirement, the pressure on women to undergo the procedures and to avoid unplanned pregnancies will be difficult to resist.

Most supporters of genetic technology do not go as far as Savulescu. They prefer to defend its use as a choice that parents should be free to make. But freedom can be justifiably limited to avoid great social ills or harm to others. A number of the contributors to this volume argue that allowing parents to 'enhance' their children through genetic therapy would cause serious harm to society or to some individuals in it.

Even Dahl, who defends the freedom of parents to choose the sex of their child, allows that this freedom should not be available in circumstances where the choices of parents are likely to result in harm: for example, in those societies where parents would mostly prefer to have boys. The ability of parents to select the characteristics of their children would open up other possibilities for harm to individuals and societies. Racial discrimination in many societies disadvantages people of minority races. To avoid these disadvantages, Black parents might choose to ensure that their children have the genetic characteristics of Whites. Cannold argues that this way of dealing with discrimination would be bad for society and especially bad for the remaining Black people because it does nothing to challenge the causes of discrimination.

Suppose that all Black parents use the technology to ensure that they have White children. Of course, this would not guarantee that their children are not subject to discrimination. Racists might continue to discriminate on the basis of ancestry, as did the Nazis. But even if parents could use genetic technology to ensure that their children would not be subject to racial discrimination, there are reasons for concern. Robert Sparrow and Mianna Lotz are worried about the way in which genetic sorting tends to identify a person with his or her genetic constitution – the very thing that is done by racists. Sparrow thinks that even existing programmes of genetic testing are problematic because they involve selecting for or against embryos on the basis of their genetic constitution. He thinks that there are morally disturbing similarities between programmes that sort fetuses according to their genetic constitution and the eugenic programmes of the past. Lotz is particularly concerned with the belief in genetic determinism which she thinks genetic technology encourages: the belief that our fate is determined by the nature of our genes.

David Neil, elaborating arguments of Habermas, maintains that programmes that allow parents to choose the characteristics of their children would undermine the very basis of a liberal democratic polity. The existence of such a polity, in his view, depends on the belief that individuals are equally worthy of respect as autonomous, self-determining beings. But if parents, or some other agent, can determine the genetic constitution of offspring, then in an important sense they cease to be self-determining. By becoming objects of manufacture they cease to be equal citizens of a democratic polity.

There is also an issue of intergenerational obligations, as Janna Thompson points out. Should parents today be restricted in their reproductive choices by reason of the effect that those choices may have on later generations? If parents are free to make whatever reproductive choices they choose, will the children born from such 'liberal genetics' ultimately be the 'best' citizens?

Most of the contributors believe that a society must put some limitation on the use of genetic technology. However, Loane Skene, in the concluding chapter, explains the difficulties in using the law to restrict the genetic tests available to parents and the circumstances in which a pregnancy may be terminated if a fetus is found to be affected. If the restricted tests are listed in legislation, the list will quickly become outdated due to the rapid development of new tests. If tests are limited to 'serious' genetic disorders, that term may be difficult to interpret. Also, in some jurisdictions (like the state of Victoria), fetal abnormality is not in itself a ground for lawful termination of pregnancy.

The issues raised in this book are of the most fundamental and challenging kind when we are deciding the kind of society in which we want to live. Readers will not agree with many of the views presented in the book but the range of views should certainly stimulate and inform discussion.

NOTES AND REFERENCES

1. A. Huxley, *Brave New World* (Harmondsworth: Penguin, 1955).

2

Genetic testing, an informed choice

Agnes Bankier and David Cram

Background

This chapter describes a range of genetic tests that are apparently available and how they offer choices to those affected. It explains some of the reasons for genetic testing and some ethical issues that have been raised concerning particular tests, especially in relation to the privacy and control of the information revealed.

Traditionally genetic counselling has aimed to be non-directive and to provide accurate information and an opportunity for discussion so that people can make informed choices for themselves. Decisions need to be based not only on scientific evidence and legal requirements but also on the personal values of the people concerned. In the context of prenatal diagnosis and predictive testing, decisions have life-long consequences for the individual and their families and need due consideration.

Gene testing is available to confirm a diagnosis in a person with symptoms of a condition and can clarify if a healthy relative or a fetus has inherited the gene in the family. Gene testing can thus provide information about the future health of the individual; this is called predictive testing. Carrier testing provides information about risk to offspring. There are screening tests, which are available to the population, for conditions of sufficient frequency that are a low known risk to individuals but can identify those who are in fact at high risk.

Prenatal testing and screening have created options for parents to find out about the health of the baby before it is born. This information can be used to prepare for the birth of the baby. In other circumstances when the baby is likely to have serious problems the parents may elect not to continue the pregnancy. Carrier and diagnostic testing is available to people who are known to be at increased risk because of their family background or on the

The Sorting Society: The Ethics of Genetic Screening and Therapy, ed. Loane Skene and Janna Thompson. Published by Cambridge University Press. © Cambridge University Press 2008.

basis of screening tests. Antenatal diagnosis has been widely practised for more than 20 years with a better than 80% uptake. These are invasive tests and involve some risk to the mother or the pregnancy. Pre-implantation genetic diagnosis (PGD) is an alternative reproductive option that has been available for more than ten years.

Ever since the beginning of the Human Genome Project, concerns were raised about the ethical, legal and social implications of genetic testing. Questions have been raised regarding how far we should go in applying these technologies, how far we should go in supporting parental choice regarding the type of children they choose to have. Questions have been asked about the definition of a serious birth defect and how the term disability is defined. Questions have been raised about the potential impact of prenatal diagnosis on the community's regard for people with disabilities, how they are valued and supported by the community and indeed the potential impact on our future society. If we regard prenatal diagnosis as a necessary part of reproductive autonomy, then it is important that all aspects of that choice are equally considered: that we are providing not only alternatives but real choices. Choices need to be supported. Supported choice is not only the right to a safe abortion and post-termination support but also support at every level for the care of the child who could be born with disabilities.

Population screening

Screening tests are available to a whole population, whose members individually are at a low known risk, with the aim of identifying those at increased risk. A screening test is not a diagnosis. Population screening may be undertaken during or before pregnancy (to identify problems in fetal development or carrier status in the parent) or after a child is born (newborn screening).

Pregnancy screening

Maternal serum screening
In maternal serum screening, participation is voluntary and involves a blood test. The screening programme will identify individuals who are at increased risk for Down syndrome or spina bifida. The risk for other chromosome abnormalities may be identified at the same time. The risk is determined according to agreed parameters, and women identified as having increased risk will have the option of more definitive testing to clarify their situation. They need the opportunity to understand the relevant information to enable them to make informed decisions.

Usually this information about the screening programme is offered in written form and genetic counselling is available to individuals who are found by screening to be at increased risk. Most of those at increased risk will still have a normal result on specific testing. It is important to ensure that individuals considering screening should understand the consequences of that decision later on, that is, that it could lead to consideration of further testing and possible termination of pregnancy.

Maternal serum screening may be offered at one of two times in the course of the pregnancy. In the first trimester a blood test at 10–12 weeks' gestation, combined with nuchal translucency ultrasound measurements at 11–13 weeks' gestation, can detect 90% of babies with Down syndrome. Maternal serum screening in the second trimester involves measurements of three metabolites (triple screen: alphafetoprotein, human chorionic gonado-trophin, unconjugated oestriol) or four metabolites (quadruple screen: alphafetoprotein, free beta human chorionic gonadotrophin, unconjugated oestriol, inhibin A) in maternal blood. The risk assessment is reported, increased risk being the top 5% of the measured value. Those at increased risk may choose to have chorionic villous sampling (CVS) or amniocentesis for a definitive diagnosis.

In Victoria 70% of pregnant women now elect to have maternal serum screening in pregnancy. Since the increase in the number of women having screening tests in pregnancy, there has been a gradual decline over the past five years in the number having invasive tests of CVS and amniocentesis. Overall more older women are having babies: compare 7.8% of women over 35 years in 1985, with 22.4% in 2004 in Victoria. The proportion of babies with Down syndrome born to younger mothers less than 34 years of age (who are much less likely to have maternal serum screening) was 70.9% in 1990 and 53.5% in 2003. The live birth prevalence of Down syndrome has not decreased significantly, as a result, despite screening and testing of older women.

Ultrasound screening

The measure of nuchal translucency in the first trimester, combined with first trimester serum screening, has been a powerful screen for chromosomal abnormalities such as Down syndrome, enabling a detection rate of up to 90%.

Carrier screening in target groups

Carrier testing can be offered to identify carriers, which in turn provides choices for them. Genes come in pairs and in a single copy may not cause signs or symptoms. In a recessively inherited condition, the person who carries a single copy of the recessive gene is healthy but can have a child born with the condition if their partner is also a carrier of the gene. All individuals carry some recessively inherited gene faults. In certain populations the likelihood of being a carrier of a particular gene is increased. Once an individual is identified as a carrier they are offered genetic counselling regarding the risks to their offspring. 'Cascade carrier testing' is then offered to the person's immediate blood relatives. That person is asked to inform family members about their potential risk and the individual is supported in doing so. One highly successful screening programme has been for Tay-Sachs disease. This worldwide initiative has virtually eliminated the condition from the Ashkenazi Jewish population, despite a carrier frequency of one in 25 in that population.

Newborn screening

Newborn screening has been established for conditions that are a significant health problem, have a known cause, a safe and reliable test, known treatment intervention and established cost-benefit. Since the late 1960s babies have been tested at a few days of age for treatable conditions in newborn screening programmes that were initially established to screen for phenylketonuria (PKU) from 1964. Testing for hypothyroidism was added in 1977 and for cystic fibrosis in 1989. In a number of centres, including all Australian states, Tandem mass spectrometry was introduced in the past decade so that some 25 additional metabolic conditions can now be screened. Screening for other disorders has been added in different centres. A blood spot, obtained by a heel prick, is tested for these conditions so that treatment can be started before the baby gets sick to prevent damage. Most of the conditions are inherited and will not be obvious at birth, even to experienced paediatricians. The blood is collected onto a special paper card, called a Guthrie card, and more recently called a newborn screening card (NBS card).

Whilst newborn screening is regarded as part of the standard of care for babies, and is clearly for the health benefit of the baby, screening is not mandated in most places and is provided as a choice to be approved by the parents. In the past decade, new genetic technologies have made it potentially possible to do gene testing on blood spots, as well as the earlier screening tests. Controversy has surrounded the storage and potential use of the blood spots for purposes other than newborn screening.

Whilst the NBS card is regarded as a health record and as such its storage and access is governed by federal and state privacy legislation, there is lack of legal clarity with respect to the status of the blood sample once it is removed from the card. The blood is no longer a health record and its use is not protected by privacy legislation or the Human Tissue Act or any other current legislation. In Victoria, access to past NBS cards is restricted by specific protocols and self-regulation. Retention of NBS card is of potential benefit to the baby, the family and to the community. Access to the blood spots is restricted to the following specific circumstances:

- **Quality assurance**: access to the blood spot is essential in case of misdiagnosis of a genetic condition. Thankfully no such case has occurred in Victoria to date, but if it did occur it would be necessary to go back to the NBS card of all babies born on that day to ensure that there were no other undiagnosed/misdiagnosed children. Access to anonymized Guthrie cards may also be necessary when the service establishes new or improved tests such as those using tandem mass spectrometry.

- **Diagnosis**: the NBS card is a convenient way of transporting blood samples for testing interstate when a particular test is not available in Victoria (this is done with parental consent). Cards may also be used for retrospective diagnosis for the child, e.g., to establish that an infection was congenital rather than occurring some time after birth. Cards may be tested for the benefit of close blood relatives of a deceased child, e.g., to establish the genetic mutation in a child who died of muscular dystrophy in order to clarify whether a sister is a genetic carrier of the mutation.
- **Forensic identification** at the request of the Coroner's Court and by court order for identification of deceased persons.
- **Research** approved by a duly constituted Hospital Research Ethics Committee (HREC). This is permitted with parental consent if identified blood samples from the NBS card are used.
- **HREC** approval is significant for using de-identified samples, such as determining the frequency of a disease-causing mutation. In such cases, access is given only to de-identified punched out blood spots, and not the NBS card itself.

Diagnostic tests

Diagnostic tests are more specific than population screening and they can also be undertaken before birth, after birth and during a patient's lifetime.

Ultrasound diagnosis

For pregnant women, ultrasound has been used to date the pregnancy, and to establish the position of the placenta and the growth of the fetus. More sophisticated ultrasound examination performed by experts – the anatomical ultrasound examination – at the appropriate time can also identify structural birth defects. Anatomical ultrasound is most informative between 18 and 20 weeks of pregnancy. In addition, ultrasound guidance has been used for invasive tests of CVS, amniocentesis or fetal biopsy.

In one study, chromosome analysis performed after ultrasound detection of a structural abnormality led to the identification of a significant chromosome abnormality in 17.4% of a group of women and in women over 37 years of age, in 25% of them. Of the major abnormalities detected through ultrasound 61.1% were associated with increased nuchal thickening.[1]

An anatomical scan between 18 and 20 weeks of pregnancy can detect many structural birth defects although some do not manifest until later in the pregnancy. The definitive diagnosis may not be made until the baby is delivered, particularly for skeletal dysplasias and brain abnormalities.

Gene tests confirming the diagnosis in a symptomatic person

Diagnostic tests are done for people of all ages, a targeted test, to confirm a diagnosis suspected because of symptoms in an individual or because of a known high risk derived from their family history or as a result of a screening test. Most diagnostic gene tests are performed on a blood sample, but a few cells obtained as a cheek brush sample may be sufficient.

Prenatal diagnosis

Prenatal diagnosis provides targeted testing to individuals of known increased risk for a specific disorder. The tests are usually performed with the aim of having a healthy baby. Whether or not people choose to have prenatal testing is a matter of personal choice.

The type of procedure undertaken depends on the stage of pregnancy, the nature of the test and personal preference. All aspects of testing and their consequences need to be discussed. Whilst testing may be performed to clarify one diagnosis, sometimes other abnormalities may be incidentally detected. If there are unexpected findings genetic counselling is provided.

These tests have been available since the 1970s but initially only for chromosomal and biochemical tests for serious disorders affecting children. Gene tests have become increasingly available in the past 15 years. Indications for prenatal diagnosis are guided by recommendations from professional bodies e.g: in Australia, the Prenatal Diagnosis Policy of the Human Genetics Society of Australasia and the Royal Australian and New Zealand College of Obstetrics and Gynaecology.[2]

Almost half of the prenatal diagnostic tests are performed because of advanced maternal age. However now over a third of tests in Victoria are done because of 'increased risk' maternal serum screening results. The overall number of invasive tests has been reduced.

Chromosome analysis, involving cell culture, takes 7 to 10 days. Fluorescent *in situ* hybridization (FISH), first introduced in 1999, can provide a chromosome count, for some of the chromosomes, within a day or two. This is achieved by viewing the number of fluorescent colour-tagged probes in a sample but only the most frequent trisomies are tested. Trisomy 21 (Down syndrome) is the most common chromosome abnormality identified. In one study, 8% of CVS and 4% amniocentesis led to the detection of a chromosome abnormality.[3]

Some cytogenetic laboratories now employ a highly accurate quantitative polymerase chain reaction (PCR) assay to detect major chromosomal abnormalities in the space of a few hours after sample reception. The main advantage of this assay is to provide a more rapid diagnosis in order to reduce the stress on patients waiting for results. It is particularly useful in situations where non-invasive ultrasound and biochemical assays predict that the fetus has a high risk of carrying a chromosome abnormality.

The first type of test for prenatal diagnosis, CVS, is performed between 10 and 15 weeks of pregnancy. A needle is inserted, under ultrasound guidance, into the uterus and a small sample is removed from the edge of the placenta. Sufficient DNA can be extracted from the sample for gene testing and biochemical analysis. The procedure carries a 1:200 chance of miscarriage. Termination, if requested, can be performed under a general anaesthetic by vaginal evacuation.

Another test, amniocentesis, is performed at 15 to 18 weeks of pregnancy. Under ultrasound guidance a needle is inserted into the amniotic sack and a sample of fluid is removed. Amniotic fluid contains cells shed by the baby, which can be tested. The cells may need to grow in culture until there are sufficient numbers for testing. This procedure carries a 1:100 chance of miscarriage. Termination of pregnancy, if requested, can be performed by inducing labour.

Other tests may be developed in future. Previous observations of fetal cells in the maternal blood and the cervical mucus during early pregnancy, together with the availability of single-cell-based tests, provide a new opportunity to perform prenatal diagnosis at an earlier stage of pregnancy (7–8 weeks). This would allow couples at genetic risk the option of early termination of pregnancy if the fetus were found to be affected. Research over the last 15 years has failed to deliver a reliable early test for prenatal diagnosis based on fetal cells from the maternal blood. This is due, in part, to the very low frequency of fetal cells in maternal blood (2–3 cells per ml), the lack of high affinity fetal cell antibodies to capture fetal cells, and the observation that these fetal cells can persist in the circulation system for many years. This could cause a misdiagnosis of the current pregnancy. Under investigation is the application of new technology to the identification and testing of fetal DNA in maternal blood.

Fetal trophoblastic cells from the cervix represent another approach for future tests because they are present at a higher frequency in the cervical mucus, and can be sampled using simple mucus aspiration techniques.[4] Improved methods have been described to reliably isolate and purify fetal cells from cervical mucus samples using specific trophoblastic cell antibodies, and confirm fetal origin by DNA fingerprinting techniques.[5] Recently, in a preclinical trial conducted by Monash IVF,[6] there was a 100% concordance between cervical fetal cells and corresponding placental tissue for sex diagnosis. Clinical trials are currently in progress to assess the safety and acceptability of cervical fetal trophoblastic cell biopsy and the accuracy of the technique for the diagnosis of genetic disease.

Pre-implantation genetic diagnosis (PGD)

Pre-implantation genetic diagnosis is performed on cells obtained from the embryo by biopsy where couples are undergoing reproduction technology for treatment. The tests provided are usually the same types of tests that are available in prenatal diagnosis. In these cases, 'PCR tests' are designed specifically to diagnose the known paternal and maternal mutations identified by prior DNA testing, or tests for known chromosome abnormalities. The couples who choose this reproductive option usually have either a moral objection to termination of pregnancy, have had one or more terminations in the past, have recently lost a child to a genetic disease or hold great hope that the new technology will deliver them an unaffected child. The aim is the transfer of unaffected embryos to the woman's uterus for implantation to help have a healthy child.

Pre-implantation genetic diagnosis is most often used where IVF has been unsuccessful, particularly for couples of advanced maternal age who have unexplained infertility. In this procedure, embryo biopsy samples are screened by FISH for numerical changes in chromosomes commonly involved in spontaneous abortions.

Worldwide, there are over 70 IVF clinics that offer a clinical PGD service with an on-site genetics laboratory. Fifteen years after the inception of PGD, in 1989, over 1000 babies had been born using the method.[7] Today, it is estimated that this figure has escalated to over 1500 babies. Comprehensive data on PGD from participating laboratories have been collected up to 2004 by the European Society of Human Reproduction and Embryology (ESHRE) PGD Consortium,[8] but unfortunately most US data are lacking. Overall, ~80% of cycles were performed by FISH for chromosomal structural abnormalities and X-linked disorders and ~20% for autosomal single gene disorders. Pre-implantation genetic diagnosis has now been performed for over 120 different single gene conditions. The most commonly requested tests for autosomal recessive conditions are for cystic fibrosis and beta thalassaemia, which is consistent with the high carrier frequency of causative mutations in the general population. Huntington disease and myotonic dystrophy are the most requested autosomal dominant conditions, whereas tests for haemophilia A and Duchenne muscular dystrophy are the most requested tests for X-linked conditions.

Prenatal diagnostic testing has proven to be a highly reliable and accurate method. For FISH and PCR, the reliability of obtaining an interpretable result is 85–90%, while the accuracy for most prenatal diagnosis tests is greater than 96%. While most clinics recommend confirmatory prenatal testing, the uptake rate ranges from 20% to 40%. It is anticipated that adoption of PGD clinical guidelines[9] across all clinics will result in incremental improvements to reliability and accuracy rates. Overall, the pregnancy rate per embryo transfer following PGD is 25%, although some clinics have reported consistently higher rates of more than 30%. Thus, even though the vast majority of PGD cycles lead to the transfer of one or two unaffected embryos, it is likely that most couples will require two or three PGD cycles to achieve a pregnancy. Occasionally, couples who have been unsuccessful in their first PGD cycle revert back to natural conception and prenatal diagnosis.

With advances in assisted reproductive technologies, such as blastocyst biopsy[10] and the development of new molecular techniques, such as multiplex fluorescent PCR, it is now possible to apply PGD to other indications beyond classical heritable disease.

Many people have fewer moral concerns about selecting one embryo rather than another for implantation, compared with abortion of a fetus in an established pregnancy. However, questions have arisen about other types of selection such as incidental sex selection, or selection against carriers. Pre-implantation genetic diagnosis has now been performed for genetic predisposition to a range of different cancers[11] and late onset Alzheimer's disease[12] where the disease risk of inheriting the mutation is less than 100%, and thus termination of pregnancy is a more difficult issue. Some patients have also used PGD to achieve a pregnancy with an embryo that is unaffected with disease and human leukocyte antigen (HLA) tissue-matched to a sick sibling (chance of 3 out of 16) so that the cord blood from the delivery can be used for stem cell transplantation. An even smaller number of patients have requested HLA matching in the absence of disease (chance of 1 out of 4) as a mean of obtaining stem cells to treat an existing child with leukaemia. Some clinics also offer social sex selection, but this practice remains controversial. In addition to these indications, PGD has also been performed, on rare occasions, for mitochondrial disease and for Y deletions that cause male infertility.[13] Due to the ethical issues associated with these requests, together with the specific IVF and PGD legislation in different countries, it is not possible for every clinic to offer PGD for all these indications. It is becoming more difficult for smaller PGD laboratories to offer a comprehensive PGD service and manage all requests. Many of these laboratories are collaborating with existing prenatal diagnosis laboratories and offer a clinical PGD service using transport of biopsy samples to an outside testing laboratory. For X-linked disease, some US laboratories also send semen samples to an external laboratory where the sample is sorted into X-bearing and Y-bearing sperm and sent back to the IVF unit where the oocytes can be fertilized with X-bearing sperm to produce female embryos with a lower risk of disease. Pre-implantation genetic diagnosis has also been successfully performed by sending a team of scientists to Cyprus where embryos produced by the local IVF clinic are tested for beta thalassaemia and sickle cell disease.

One of the major deficiencies of the FISH technique is that only a limited set of chromosomes (i.e. X, Y, 13, 14, 15, 16, 18 and 21) can be screened in a single cell. Thus, it is likely that embryos with other chromosomal aneuploidies are still being transferred to the patient. While most of these aneuploidies are lethal prior to, or at the time of implantation, pregnancy rates are clearly being compromised. Although the technique of comparative genome hybridization (CGH) using metaphase chromosomes or microarray probes as a read out for total chromosomal aneuploidy holds promise, it has yet to be developed and validated sufficiently for clinical application. The development of a reliable CGH-based technique would also allow combined aneuploidy and single gene testing to ensure that disease free embryos are also chromosomally normal. The availability of blastocyst biopsy to take a larger cell sample of 10–15 trophectoderm cells for testing may also allow the identification of gene expression patterns associated with viable embryos and new tests to predict embryo viability prior to transfer. This would allow the transfer of a single viable embryo, increasing pregnancy rates and reducing the incidence of multiple gestations. Finally, the new technology may allow the identification of imprinting disorders that have been associated more often with IVF procedures.[14]

Some proponents of PGD believe that all IVF embryos in the future should be tested for aneuploidies given that aneuploidy rates in eggs from younger women are around 20%, and that the incidence of embryonic chromosomal mosaicism is high in human IVF embryos. Whilst embryo biopsy at the cleavage or blastocyst stage appears to be a safe procedure with over 1000 babies born to date, long-term follow-up of such children is needed to determine if there are any adverse effects of technology. Dissemination of scientific information related to PGD outcomes and new technologies will continue to play an important role, so that clinicians, genetic counsellors and patients are fully aware of the implications of new PGD tests.

Predictive testing

Since we are born with all our genes and gene faults, it is possible to do genetic testing at any point in time. Individuals known to be at increased risk because of their family history, such as Huntington disease or breast cancer, can be tested to establish whether or not they have inherited the gene before they develop any signs of the condition. The person who is identified to have inherited the gene is healthy at the time but is aware of the prospect of developing the disorder in his/her lifetime. Since many of these genes have incomplete penetrance, not all of people who inherit the gene will develop the condition. The nature of the mutation, or gene change, must first be established in an individual who does have the condition to ensure that the detected gene change is indeed causative. Predictive testing should only be undertaken after appropriate genetic counselling to those given the opportunity to consider the pros and cons of being tested. Though gene or other therapy is not an option in most cases, heightened surveillance strategies can improve health outcomes for those who develop the condition.

Predictive testing is not recommended in minors if no treatment interventions are available which would alter the course of the condition.

Cascade testing

When a genetic condition occurs in the family, gene testing can establish who in the lineage has inherited the gene. Carrier testing in minors is not recommended but can be considered on a case-by-case basis in adolescents.

Future tests

Genes chips, MLPA (multiple ligation-dependent probe amplification), application of tandem mass spectrometry to detect single base changes and other new technologies will generate more detailed, faster and cheaper genetic tests which will identify genetic cause and predict susceptibilities to complex disorders. Genetic profiling will predict the likelihood of developing a disorder and provide the opportunity for therapeutic and environmental interventions. Other tests will help to clarify the drug treatment most suitable for the individual according to their genetic profile. This developing area of pharmacogenomics will help avoid the toxic side effect of drugs, which currently create a significant burden in health care. It will also make it possible to use some drugs in an effective and targeted way, including drugs currently not used because of toxic effects to susceptible individuals. The health benefit and cost benefit of these tests will need considered evaluation.

Concerns have been raised regarding cloning, stem cell therapies and potential use of genetic testing to select babies not only on the basis of health but also for desired physical characteristics. Apart from sex selection, tests for specific physical characteristics are not available. The determinants of characteristics such as height and intelligence are complex, determined by the combination of many genes as well as environmental factors. Whilst there are concerns that PGD could be used to produce 'designer babies', in reality the complexity of the determinants of traits and of PGD in the setting of assisted reproduction severely limits the potential of such options.

Stem cell research is progressing but is limited by the availability of stem cells. Disease-specific stem cells, harvested from embryos rejected by PGD, could provide an invaluable source of cells for research and for trial therapies. Reproductive cloning is complex and has medical risks documented from animal research and is not legal or supported even in principle, internationally.

NOTES AND REFERENCES

1. E. Muggli and J. Halliday, (2003) Report on Prenatal Diagnostic Testing in Victoria 2002, www.health.vic.gov.au/perinatal/downloads/report_diagnostictest 2002.pdf; 18.

2. Royal Australian and New Zealand College of Obstetrics and Gynaecology, www. hgsa.com.au.

3. *Ibid.*

4. R. Cioni, C. Bussani, B. Scarselli *et al.*, 'Fetal cells in cervical mucus in the first trimester of pregnancy' (2003) 23 *Prenatal Diagnosis* 168–71.

5. J. N. Bulmer, R. Cioni, C. Bussani *et al.*, 'HLA-G positive trophoblastic cells in transcervical samples and their isolation and analysis by laser microdissection and QF-PCR' (2003) 23 *Prenatal Diagnosis* 34–9; M.G. Katz-Jaffe, D. Mantzaris and D. S. Cram, 'DNA identification of fetal cells isolated from cervical mucus: potential for early non-invasive prenatal diagnosis' (2005) 112 *BJOG* 595–600.

6. D. Mantzaris, D. S. Cram, C. Healy, D. Howlett and G. Kovacs, 'Preliminary report: correct diagnosis of sex in fetal cells isolated from cervical mucus during early pregnancy' (2005) 45 *Australian & New Zealand Journal of Obstetrics & Gynaecology* 529–32.

7. Y. Verlinsky, J. Cohen, S. Munne *et al.*, 'Over a decade of experience with preimplantation genetic diagnosis: a multicenter report' (2004) 82 *Fertility and Sterility* 292–4.

8. J.C. Harper, K. Boelaert, J. Geraedts *et al.*, 'ESHRE PGD Consortium data collection V: cycles from January to December 2002 with pregnancy follow-up to October 2003' (2006) 21 *Human Reproduction* 3–21.

9. A. R. Thornhill, C. E. deDie-Smulders, J. P. Geraedts *et al.*, 'ESHRE PGD Consortium 'Best practice guidelines for clinical preimplantation genetic diagnosis (PGD) and preimplantation genetic screening (PGS)" (2005) 20 *Human Reproduction* 35–48; The Preimplantation Genetic Diagnosis International Society (PGDIS), 'Guidelines for good practice in PGD' (2004) 9(4) *Reproductive Biomedicine Online* 430–4 (PGDIS (2004)).

10. K. A. de Boer, J. W. Catt, R. P. Jansen, D. Leigh, S. McArthur, 'Moving to blastocyst biopsy for pre-implantation genetic diagnosis and single embryo transfer at Sydney IVF' (2004) 82(2) *Fertility and Sterility* 295–8; G. Kokkali, C. Vrettou, J. Traeger-Synodinos *et al.*, 'Birth of a healthy infant following trophectoderm biopsy from blastocysts for PGD of beta-thalassaemia major' (2005) 20(7) *Human Reproduction* 1855–9. Epub 2005 May 5.

11. Y. Verlinsky, S. Rechitsky, O. Verlinsky *et al.*, 'Preimplantation diagnosis for P53 tumour suppressor gene mutations' (2001) 2(2) *Reproductive Biomedicine Online* 102–5.

12. Y. Verlinsky, S. Rechitsky, O. Verlinsky, *et al.*, 'Preimplantation diagnosis for early-onset Alzheimer disease caused by V717L mutation' (2002) 287 *JAMA* 1018–21.

13. J. C. Harper, K. Boelaert, J. Geraedts *et al.*, 'ESHRE PGD Consortium data collection V: cycles from January to December 2002 with pregnancy follow-up to October 2003', note 8 above.

14. C. Allen and W. Reardon, 'Assisted reproduction technology and defects of genomic imprinting' (2005) 112 *BJOG* 1589–94.

3

Sex selection: sorting sperm as a gateway to the sorting society?

Edgar Dahl

Since ancient times, couples have been trying to influence the sex of their children. Following a suggestion by Aristotle, they were making love in the north wind to ensure the birth of a son and in the south wind to ensure the birth of a daughter. According to a proposal made by Galen, men were tying a string around their left testicle to make a boy and tying the right one to make a girl. In medieval times the proposed formula got even more bizarre when alchemists recommended to drink the blood of a lion and then have intercourse under a full moon to sire a son.[1]

Choosing the sex of our children is no longer a fantasy. However, the prospect of a reliable method for sex selection has not only raised old hopes, but also new fears. Many people are concerned that it may lead to an imbalance of the sexes, most likely a preponderance of males. Such an overabundance of men and a shortage of women, some sociologists have predicted, will invariably cause an enormous rise in enforced celibacy, polyandry, homosexuality, prostitution, rape and other sexual crimes.[2] Many feminists are similarly alarmed. Some have called the deliberate choice of a male child 'the original sexist sin'.[3] Others went so far as to warn us of an impending 'gynocide'.[4] Are these fears justified? How well are they supported by empirical evidence? And most of all: does sex selection call for a legal ban?

Before turning to these questions, let us briefly review the state of the art. Currently, there are three different types of sex selection: sex-selective abortion, sex-selective embryo transfer and sex-selective insemination. Sex-selective abortion has been made possible by prenatal diagnosis. Amniocentesis, chorionic villus sampling and ultrasound not only allow for the detection of fetal abnormalities, but also for the determination of fetal sex. In principle, women may use the information to decide whether or not to terminate a pregnancy if the fetus is not of the desired sex. Using prenatal diagnosis for the

The Sorting Society: The Ethics of Genetic Screening and Therapy, ed. Loane Skene and Janna Thompson. Published by Cambridge University Press. © Cambridge University Press 2008.

sole purpose of sex-selective abortions is, however, very rare in Western societies. For example, a follow-up study of 578 patients having prenatal diagnosis at one Melbourne centre found that none of the women had a termination for fetal sex.[5] Going through the traumatizing experience of an abortion is usually seen as too high a price for a child of a particular sex.

Sex-selective embryo transfer has been facilitated by the arrival of pre-implantation genetic diagnosis,[6] which is an alternative to prenatal diagnosis. It offers couples who are at risk of transmitting a genetic defect and are having in-vitro fertilization, the opportunity to have their embryos screened before they are transferred into the uterus. Since only those embryos that are free of the abnormality concerned will qualify for transfer to the womb, pre-implantation genetic diagnosis reduces the risk of bearing a child with a genetic disease and helps to avoid the difficult decision whether or not to terminate a pregnancy. Like prenatal diagnosis, pre-implantation genetic diagnosis can also be used to determine the sex of the embryos. Thus, women may request to transfer only those embryos that are of the desired sex.

Sex-selective insemination has become possible with the recent development of a sperm separation technology called MicroSort, which relies on an identifiable difference between X and Y chromosome-bearing spermatozoa. X- and Y-bearing sperm cells differ in their total DNA content by 2.8%, owing to the larger size of the X chromosome. A flow cytometric separation yields an average of 92% X-bearing and 73% Y-bearing sperm populations.[7] The separated sperm populations can then be used for intrauterine insemination. Given that not every attempt of artificial insemination results in a pregnancy, couples will have to undergo an average of three to five cycles of insemination. Each attempt will cost about £1250.

It can seem, therefore, that there is not yet any convenient method of sex selection. Sex-selective abortion requires the termination of a pregnancy; sex-selective embryo transfer necessitates in-vitro fertilization; and sex-selective insemination is still ineffective. However, since it is likely that MicroSort will soon be refined, sex-selective insemination is certainly the technology of the future. As soon as this sperm separation technology develops into a safe and reliable procedure, sex selection may become attractive to many couples. The only thing that would be needed to have a child of the preferred sex would be a visit to a fertility clinic and several cycles of intrauterine insemination.

Scientific interest in the development of a sperm separation technique has mainly arisen from the desire to prevent X-linked disorders. There are more than 500 sex-linked diseases in humans, including haemophilia, Duchenne's muscular dystrophy, Lesch-Nyhan syndrome and Tay-Sachs disease.[8] In most cases, the X-linked disorders are only expressed in the male offspring of carrier mothers. Thus, women who are carriers of a severe sex-linked disease often choose to have no children or to terminate their pregnancy if prenatal testing reveals the fetus to be a boy. A reliable sperm separation technology such as MicroSort would allow for the exclusive conception of unaffected girls.

Sex selection for the prevention of X-linked disorders is generally regarded as morally acceptable. The public debate, therefore, focuses almost entirely on sex selection for non-medical reasons. Many European countries, including Austria, Switzerland, Belgium, Italy and Germany, have passed legislation that makes sex selection for any but the most serious of medical reasons a crime. For instance, Germany's notorious Embryo Protection Act 1990 considers social sex selection a criminal offence punishable by one year of imprisonment. In the Australian state of Victoria, the sentence is even harsher. Under section 50 of the Infertility Treatment Act 1995 (Vic) doctors performing sex selection for non-medical reasons face up to two years of imprisonment in addition to '240 penalty units', which is equivalent to a fine of $A24 000. More recently, Canada enacted the Assisted Human Reproduction Act 2004 declaring that doctors performing sex selection for non-medical reasons are guilty of 'an offence and liable, on conviction on indictment, to a fine not exceeding $CAN500 000 or to imprisonment for a term not exceeding ten years, or to both.'

Is there any valid justification for criminalizing social sex selection and for sentencing a doctor to jail for, say, helping the parents of three boys to finally conceive a girl? I don't think so – at least not in a Western liberal democracy.

Western societies are pluralistic societies. They consist of individuals with different concepts of the meaning of life, of the existence of God and of the ways to pursue happiness. Consequently, in modern societies there will always be irresolvable differences over what is good for human beings. If a government tries to impose a particular morality upon its citizens, social conflict is inevitable. To avoid social tension and to deal with the moral pluralism of its citizens, Western society ought to be based upon a 'presumption in favour of liberty': each citizen should have the right to live his or her life as he or she chooses so long as they do not infringe upon the rights of others. The state may interfere with the free choices of its citizens only to prevent harm to others.

This so-called 'harm principle', which has been developed by Wilhelm von Humboldt and John Stuart Mill, has three important implications. Firstly, the burden of proof is always on those who opt for a legal prohibition of a particular action. It is they who must show that the action in question is going to harm others. Secondly, the evidence for the harms to occur has to be clear and persuasive. It must not be based upon highly speculative sociological or psychological assumptions. And thirdly, the mere fact that an action may be seen by some as contrary to their moral or religious beliefs does not justify a legal prohibition. The domain of the law is not the enforcement of morality, but the prevention of harm to others.[9]

With this in mind, we shall turn to some of the most common objections to social sex selection and inquire whether they provide a rational basis for outlawing it.

A constantly recurring objection to sex selection is that choosing the sex of our children is to 'play God'. This religious objection has been made to all kinds of medical innovations. For example, using chloroform to relieve the pain of childbirth was once considered contrary to the will of God as it avoided the 'primeval curse on woman'. Similarly, the use of inoculations was opposed with sermons preaching that diseases are 'sent by Providence' for the punishment of sin and it is wrong of man to escape from such divine retribution. Since even fundamentalist Christians ceased to regard the alleviation of pain and the curing of diseases as morally impermissible, it is hard to take this objection seriously. What was once seen as 'playing God' is now seen as acceptable medical practice. More importantly, the objection that sex selection is a violation of 'God's Law' is an explicit religious claim. As Western liberal democracies are based on a strict separation of state and church, no government is entitled to pass a law to enforce compliance with a specific religion. People who consider the option of sex selection as contrary to their religious belief are free to refrain from it, but they are not permitted to use the coercive power of the law to impose their theology upon all those who do not share their religious world view.[10]

Some are opposed to sex selection because they have the feeling it is somehow 'unnatural'. Like the objection that choosing the sex of our children is playing God, the claim that sex selection is not natural most often expresses an intuitive reaction rather than a clearly reasoned moral response. That a particular human action is unnatural in no way implies that it is morally wrong. To transplant a heart to save a human life is certainly unnatural, but is it for that reason immoral? Surely not! Thus, if we have to decide whether an action is morally right or wrong we cannot settle the issue by asking whether it is natural or unnatural.[11]

A more serious objection to sex selection is based on the claim that medical procedures ought to be employed for medical purposes only. Flow cytometric sperm separation, it is argued, is a medical technology designed to enable couples who are at risk of transmitting a severe sex-linked genetic disorder to have a healthy child. In the absence of a known risk to transmit a serious X-linked disease, there is simply no valid justification for employing flow cytometric sperm separation. This is a familiar objection in debates over novel applications of genetic and reproductive technologies. However, familiar as it may be, it is certainly not a persuasive one. We have already become accustomed to a healthcare system in which physicians provide services that have no direct medical benefit but that do have great personal value for the individuals seeking it. Given the acceptance of breast enlargements, hair replacements, ultrasound-assisted liposuctions and other forms of cosmetic surgery, one cannot, without calling that system into question, condemn a practice merely because it uses a medical procedure for lifestyle or child-rearing choices.[12]

A related objection is that offering a service for social sex selection constitutes an inappropriate use of limited medical resources. Again, if offering face-lifts is not considered to be a misallocation of scarce medical resources, it is hard to see how offering sex selection can be considered a misallocation of scarce medical resources. Moreover, by implying that every time a patient gets a nose-job another patient misses out on a bypass, this objection betrays a severely distorted conception of economics. If at all, this argument may apply to a state-run socialist economy based on a five-year plan, but certainly not to a private-run capitalist economy based on a free market. A chef opening a restaurant offering French cuisine does not deprive us of our daily bread. Similarly, a doctor opening up a fertility centre offering sex selection does not deprive us of basic health care. Provided their businesses are set up privately and their services are paid for privately, they don't deprive anyone of services.

Perhaps the most powerful objection to sex selection is that it may distort the natural sex ratio and lead to a socially disruptive imbalance of the sexes, as has occurred in countries such as India and China.[13] However, it is an empirical question whether a distortion of the natural sex ratio poses a real threat to Western societies. It cannot be answered by intuition, but only by scientific evidence. For a severe sex ratio distortion to occur, at least two conditions must be met. Firstly, there must be a significant preference for children of a particular sex, and secondly there must be a considerable demand for a reproductive service for social sex selection. It is important to note that both conditions need to be met simultaneously. For example, if there was a marked preference for children of a particular sex, but couples were unwilling to use sex selection technology (because it was thought to be too intrusive, too expensive, immoral or simply against their religion), then a readily available service for sex selection would not have any demographic effect. Likewise, if there was considerable interest in employing sex selection technology, but couples did not have a marked preference for children of a particular sex (because they wish to have an equal number of boys and girls), then, again, a readily available service for preconception sex selection would not alter the sex ratio in any way.[14]

In order to ascertain whether the two preconditions for a sex ratio distortion are indeed met, we have conducted representative social surveys in Germany, the UK and the USA. In all three surveys, a randomized, computer-assisted telephone interview tool was utilized to ask more than 1000 men and women between the age of 18 and 45 about their gender preferences and their interest in employing sex selection through sperm sorting.[15]

The first question asked was: 'If given a choice would you like your first-born child to be a boy or a girl?' In Germany, 14% of respondents preferred their first-born child to be a boy, 10% preferred it to be a girl, and a vast majority of 76% stated that they simply did not care about the sex of their first-born baby. The UK survey yielded a similar result: 16% of British men and women preferred their first-born child to be a boy, 10% a girl, and 74% said they did not care. In the USA, however, respondents had much stronger gender preferences for their first-born child: 39% preferred a boy, 19% preferred a girl, and only 42% did not mind the sex of their first child.

The second question was: 'If you were to have more than one child, would you prefer to have only boys, only girls, more boys than girls, more girls than boys, an equal number of boys and girls, or does the sex of your children not matter to you?' In Germany, 1% said they would like only boys, 1% only girls, 4% more boys than girls, 3% more girls than boys, 30% an equal number of boys and girls, and 58% stated that they did not care. In the UK, 3% wanted only boys, 2% only girls, 6% more boys than girls, 4% more girls than boys, an astonishing 68% an equal number of boys and girls, and 16% did not care about the sex of their offspring. In the USA, 5% stated they would like only boys, 4% only girls, 7% more boys than girls, 6% more girls than boys, 50% an equal number of boys and girls, and 27% said they did not mind their children's sex.

The third question inquired about the participants' interest in using MicroSort. In order to make an informed decision, they were told what this technology actually implies. Thus, they were informed that they would have to visit a fertility centre, to provide a sperm sample for flow cytometric separation, to undergo an average of three to five cycles of intrauterine insemination, and to pay a fee of approximately £1250 per attempt. In Germany, 6% could imagine taking advantage of MicroSort; however, an overwhelming majority of 92% found it simply to be out of the question. In the USA, the response was quite similar to that in Germany: only 8% could conceive of employing MicroSort, 18% were undecided, and 74% said they would not want to use it. Interestingly, in the UK, 21% were responsive to the idea of using MicroSort, 7% were undecided, and 71% said they cannot imagine taking advantage of it.

To establish whether the 92% of Germans and the 74% of Americans who rejected the idea of using MicroSort were in fact not interested in selecting the sex of their offspring or simply found the procedure to be too demanding, they were asked: 'Suppose the technology would require just a single cycle of artificial insemination, could be performed in any doctor's office and would be covered by your health insurance, would you then consider taking advantage of it?' Given these less demanding circumstances, 5% of Germans and 12% of Americans were prepared to reconsider their unwillingness to use MicroSort, while 94% of Germans and 64% of Americans still rejected the idea of using it; 1% of Germans and 24% of Americans stated they were not sure.

Finally, we asked the participants to imagine that there was a medication to select the sex of their children. Rather than visiting a fertility clinic, they would simply have to take a 'blue pill' to ensure the birth of a boy or a 'pink pill' to ensure the birth of a girl. While 8% of Germans and 18% of Americans were willing to use such a medication, 90% of Germans and 60% of Americans would not want to do so; 2% of Germans and 22% of Americans were undecided. (Data from our UK survey have not yet been analysed.)

We are not sure why only 6% of Germans, 8% of Americans and 21% of Britons are interested in sex selection through sperm sorting. However, an additional German survey suggests a plausible answer. When we asked 1005 men and women about their moral attitudes towards sex selection and whether or not it should be made available to all couples requesting it, 86% of Germans were strongly opposed to it. In order to identify the concerns underlying the widespread enmity to social sex selection, participants were then asked about the reasons for their opposition. Eighty-seven per cent of respondents said that 'children are a gift and deserve to be loved for what they are, regardless of any characteristics such as beauty, intelligence or sex'; 79% claimed that sex selection is 'playing God'; 76% were opposed because it was seen as 'unnatural'; 49% were afraid that it might skew the sex ratio; and 40% considered it to be 'sexist'. In light of this survey, it is safe to assume that the lack of interest in preconception sex selection is largely due to the fact that the overwhelming majority of Germans are strongly opposed to it.[16]

The same explanation might apply to the lack of interest in sex selection observed in the UK and the USA. According to a MORI opinion poll commissioned by the Human Fertilization and Embryology Authority, '69% of Britains do not agree with the liberal proposition that any parent should have the right to choose the sex of their child.'[17] Similarly, according to a nationwide social survey commissioned by the Genetics & Public Policy Center at the Johns Hopkins University, 'two thirds of Americans disapprove of sex selection for non-medical reasons'.[18]

Asking the proverbial 'man [or woman] on the Clapham omnibus' whether or not he would like to take advantage of MicroSort might not be the best way to determine the actual interest in using sex selection. After all, for most people it is a hypothetical question. Thus, we decided to survey pregnant women before having their first ultrasound. Do pregnant women (who usually spend some time wondering whether they are going to have a boy or a girl) have stronger gender preferences than the general public? Are they more interested in employing sex selection technology? And do their moral attitudes differ from the public at large?

In Germany, pregnant women do indeed differ from the general population. Firstly, they do have a significantly stronger preference for girls. While only 10% of the population at large wished their first-born child to be a girl, 18% of pregnant women hoped to have a baby girl first. Secondly, and quite surprisingly, they are even less interested in employing sex selection technology. Whereas 6% of the general population could imagine using MicroSort, only 3% of pregnant women could. And thirdly, their moral attitudes towards sex selection are much more 'liberal'. While 86% of the public at large were in favour of outlawing sex selection, only 57% of pregnant women were in support of a legal ban; 27% were undecided; and 16% were actually in favour of offering sex selection to all couples requesting it.[19]

A US survey conducted at Cleveland State University yielded a similar result. Of 140 women who were pregnant for their first time, 18% preferred to have a boy, 23% preferred to have a girl, and 59% expressed no preference at all. Asked 'If the means were available to you so that you could have selected the sex of your child, would you have done so?', 18% answered with yes, 53% with no, and 29% were undecided. Of the 26 women who said they would have used sex selection, 13 would have done so to ensure the birth of a boy and 13 would have done so to ensure the birth of a girl.[20]

Finally, a UK survey conducted at the Centre for Family Research of the University of Cambridge produced a result akin to that of Germany and the USA. Of 2359 pregnant women who had been asked 'Do you mind what sex your baby is?', 6% preferred a boy, 6% preferred a girl, 12% quite liked a boy, 19% quite liked a girl, and 58% said they had no preference for a child of a particular sex.[21]

As we know all too well, there is often a yawning gap between what people say and what they actually do. Thus, it is reassuring that demographic research that has focused on examining when couples stop having more children does indeed confirm the stated preference for a so-called 'gender balanced family'. Couples with two boys and couples with two girls are more likely to have a third child than couples with one boy and one girl – suggesting that parents with children of both sexes are more content with their family composition. This distinct trend towards a balanced family has not only been observed in Germany, the UK and the USA, but also in Canada, Italy, Spain, Sweden, Belgium, Austria, Switzerland and the Netherlands.[22]

Maybe even more instructive than social surveys are data published by so-called 'gender clinics'. Worldwide, there are about 75 fertility centres that offer some method of sperm sorting followed by intrauterine insemination. According to The London Gender Clinic, within its first 18 months it had been consulted by only 809 couples. Of the 809 couples, 468 were of Indian origin, 259 European, 29 Chinese and the remaining 53 of other ethnic origins. The majority of European couples were seeking sex selection to 'balance their family', i.e. they already had two or three children of the same sex and wanted to have at least one child of the opposite sex: 'Our study shows that well over 95% of couples came for this sole purpose. They are predominantly men and women in their mid-30s nearing the end of their reproductive life and having on average 2-3 children of the same sex'.[23] Similarly, the Gender Clinic of New York City reports that all of the 120 American couples seeking sex selection were doing so for family balancing purposes: 'They selected girls when they had boys at home and boys when there were only girls.'[24] Likewise, Gametrics Limited in Alzada, Montana, which detailed the collective experience of 65 gender clinics says: 'The overwhelming majority had two or more children of the same sex and desired a child of the opposite sex.[25] And finally, a report of the Genetics & IVF Institute in Fairfax, Virginia, which is currently conducting a clinical trial on the safety and efficacy of MicroSort, states: 'The majority of couples (90.5%) in our study were seeking gender preselection for family balancing purposes, were in their mid-thirties, had two or three children of the same sex, and desired only one more child.'[26]

In conclusion, the widespread fear of a sex ratio distortion seems to be unjustified. The existing empirical evidence suggests that a readily available service for preconception sex selection will have only a negligible societal impact and is highly unlikely to cause a severe imbalance of the sexes in Western societies.

Although the threat of a sex ratio distortion is potentially the most troubling problem, it is also a problem that is most easily resolved – namely by limiting the service for sex selection to the purpose of 'family balancing'. If access to sex selection were restricted to parents having at least two children of the same sex, then helping them to have a child of the opposite sex would, if at all, only marginally alter the balance of the sexes.[27]

While a severe distortion of the natural sex ratio may not be a problem in Western countries, it surely is a problem in some Asian countries such as Pakistan, India, China and Korea. In India, thousands of girls are aborted, abandoned, neglected or even killed right after birth. The introduction of prenatal testing and selective abortion has apparently skewed the sex ratio of some regions of India to such an extent that there are now only 793 girls for every 1000 boys. According to a recent survey, 'prenatal sex determination and selective abortion accounts for half a million missing female births yearly.'[28] Given that the practice of sex-selective abortions has been common for most of the past two decades, it seems that about ten million female babies might have been aborted in India alone. In February 2003, the Indian parliament took action by amending its 'Preconception and Prenatal Diagnostic Techniques (Prohibition of Sex Selection) Act of 1994'. Doctors violating the Act now face a prison sentence of up to five years or a fine of 10 000 to 50 000 Rupees. Despite increased efforts to enforce the Act, however, the practice of sex-selective abortions seems to be continuing.[29]

There are religious as well as economic reasons why Indians prefer boys over girls. According to Hinduism, a man who has failed to sire a son cannot achieve salvation. Only a male descendant can perform the last funeral rites to ensure the redemption of the departed soul. More importantly, Indian custom is that the parents of a girl are expected to pay a dowry for her marriage. The dowry payments are considerable. They extend from £2500 to £75 000. To marry off one or more daughters is therefore a huge financial burden. Since girls are a liability and boys are an asset, Indian couples have a strong incentive for seeking sex-selective abortions. Consequently, many medical practitioners offering ultrasound scans for sex determination have taken advantage of the excessive dowry demands when advertizing their services with the slogan 'Invest 500 Rupees now, save 500 000 Rupees later!'[30]

Given the number of female abortions in India, it does not come as a surprise that some authors have called for a worldwide ban on social sex selection.[31] However, does the practice of social sex selection in India really justify prohibiting social sex selection in countries such as Germany, the UK or the USA? The simple answer is: most certainly not! Firstly, preventing, say, British couples from choosing the sex of their children will not change the sex ratio of India. Secondly, even if it is only meant to 'send a message', it is simply naive to assume that Indian families will appreciate our gesture, well-meaning as it may be. As long as there are religious and economical incentives for preferring boys over girls, our moral plea will fall on deaf ears on the subcontinent. Thirdly, legalizing social sex selection in Great Britain does not jeopardize our right to criticize the practice of social sex selection in India. Approving of social sex selection through cytometric sperm separation or pre-implantation genetic diagnosis in no way implies that we approve of social sex selection through abortion or infanticide. Fourthly, and most importantly, denying British couples the opportunity to have a daughter because Indian couples have killed their girls would amount to punishing the innocent. There is no moral justification whatsoever for punishing the people of one country for actions committed by the people of another.[32]

As we have seen, Western societies do not have to worry about an impending imbalance of the sexes. However, even if social sex selection may not distort the natural sex ratio, it may distort the natural birth order. Given that a substantial portion of the US population still prefers their first child to be a boy, it could be argued that sex selection technology may be abused to ensure the birth of a first-born 'son and heir'. Hence, another objection to social sex selection claims that we are at risk of 'creating a society of little sisters.' Once girls are second born, they will feel only second best.[33] Is this a sound objection? I do not think so. Although it is, quite literally, 'conceivable', it is highly unlikely that hundreds of couples would employ sex selection technology for their first child to be a boy. As already pointed out, data from American and British 'gender clinics' suggest that the only couples willing to subject themselves to an intrusive and expensive treatment for preconception sex selection are couples who already have two or three children of the same sex and long to have at least one child of the opposite sex. According to a study by Nan P. Chico who analysed 2505 letters from couples inquiring about sex selection, 'only 1.4 per cent of these couples were seeking a first-born son, contrary to expectations based on an extensive review of the literature on gender preferences.'[34] Moreover, there is nothing in the literature on birth order studies that could possibly justify the claim that children who are second born feel second best.[35]

Another frequently advanced objection claims that sex selection is 'inherently sexist'. For example, the feminist philosopher Tabitha Powledge argues that, 'we should not choose the sexes of our children because to do so is one of the most stupendously sexist acts in which it is possible to engage. It is the original sexist sin.' Sex selection, she continues, is deeply wrong because it makes 'the most basic judgement about the worth of a human being rest first and foremost on its sex.'[36] However, this argument is deeply flawed. It is simply false that all people who would like to choose the sex of their children are motivated by the sexist belief that one sex is more valuable than the other. As we have seen, almost all couples seeking sex selection are simply motivated by the desire to have at least one child of each sex. If this desire is based on any beliefs at all, it is based on the quite defensible assumption that raising a girl is different from raising a boy, but certainly not on the preposterous idea that one sex is 'superior' to the other.

A further common objection concerns the welfare of children born as a result of sex selection. Thus, it has been argued that sex-selected children may be expected to behave in certain gender-specific ways and risk being resented if they fail to do so. Although this cannot be completely ruled out, it is highly unlikely that children conceived after MicroSort are going to suffer from unreasonable parental expectations. Couples seeking sex selection to ensure the birth of a daughter are well aware that they can expect a girl, not some Angelina Jolie; and couples going for a son know perfectly well they can expect a boy, not some Brad Pitt.

Given that social sex selection is going to be a private reproductive service that needs to be paid out of pocket, some may worry about 'social injustice' arising from unequal access to technology. Hence it could be argued: 'Allowing wealthy couples to employ yet another highly expensive medical procedure would widen the gap between the rich and the poor beyond all bearing. No one should be allowed to use this technology unless all members of our society have access to it!' I do not wish to make fun of a serious ethical problem, but I cannot help being reminded of Woody Allen's famous joke in 'Annie Hall' where two old ladies are dining in an expensive restaurant in the Catskills: 'The food in this place is really terrible', says the first one. 'Yes', agrees the other, 'and such small portions!' The point is, I gather, sufficiently clear. If something is bad, we have no reason to complain that there is not enough of it. A complaint only makes sense if we are talking about some good. The same goes for sex selection. Opponents of social sex selection are undermining their own stance if they are criticizing the practice because it will inadvertently lead to inequality of access. If they are concerned about social justice they should not oppose sex selection, but promote it by claiming coverage through their public health insurance. As to the objection that nobody should use a new technology unless there is a guarantee that its benefits are equally shared by all members of society, this is, once again, a demand that ignores the economic facts of life. When first developed, automobiles, television sets and personal computers were so expensive only a minority could afford buying them. The only reason virtually all of us can now afford to buy a car, a TV or a PC is that we did not interfere with the creation of a market, stimulating competition and, consequently, lowering the prices. Thus, if we are really concerned about poor couples being 'left out', we should encourage, not discourage, couples to employ sex selection technology. This is potentially the best way to lower its current price.

Last but not least, there is the widely popular objection that sex selection is the first step down a road that will inevitably lead to the creation of 'designer babies'. Once we allow parents to choose the sex of their children, we will soon find ourselves allowing them to choose their eye colour, their height, or their intelligence. This slippery slope objection calls for three remarks. Firstly, it is not an argument against sex selection per se, but only against its alleged consequences. Secondly, and more importantly, it is based on the assumption that we are simply incapable of preventing the alleged consequences from happening. However, this view is utterly untenable. It is perfectly possible to draw a legal line permitting some forms of selection and prohibiting others. Thus, if selection for sex is morally acceptable but selection for, say, intelligence is not, the former can be allowed and the latter not. And thirdly, the slippery slope argument presumes that sliding down the slope is going to have detrimental, if not devastating, social effects. However, in the case of selecting offspring traits this is far from obvious. What is so terrifying about the idea that some parents may be foolish enough to spend their hard-earned money on genetic technologies just to ensure their child will be born with big brown eyes or black curly hair? I am sorry, but I cannot see that this would herald the end of civilization as we know it.

Since it cannot be established that preconception sex selection for non-medical reasons would cause any harm to others, a legal ban is ethically unjustified. However, the fact that social sex selection ought not to be prohibited does not preclude regulating its practice. For example, it seems entirely appropriate to limit sex selection services to licensed centres subject to monitoring by health authorities.[37] This would not only guarantee high scientific standards and high quality professional care, but it would also enable detailed research on possible demographic consequences and thus allow action if, contrary to expectations, significant imbalances were to develop.[38]

NOTES AND REFERENCES

1. D. S. Davis, *Genetic Dilemmas: Reproductive Technology, Parental Choices, and Children's Futures* (New York: Routledge, 2001).
2. G. Vines, 'The hidden cost of sex selection' (1993 May 1) 138(1871) *New Scientist* 12.
3. T. Powledge, 'Unnatural selection: on choosing children's sex' in H. B. Holmes, B. B. Hoskins and M. Gross (eds.), *The Custom-Made Child? Women-Centered Perspectives* (Totowa, New Jersey: Humana Press, 1981) 197.
4. J. J. Raymond, *Women as Wombs. Reproductive Technology and the Battle Over Women's Freedom* (San Francisco: Harper, 1993).
5. L. de Crespigny, H. P. Robinson, A. C. Ngu, A. Robertson and J. L. Halliday, 'Transabdominal chorionic villus sampling: a safe and reliable procedure' (1991) 31(22) *Australian & New Zealand Journal of Obstetrics & Gynaecology*, 22–5.
6. A. Kuliev and Y. Verlinsky, 'Current features of preimplantation genetic diagnosis' (2002) 5 *Reproductive Biomedicine Online* 294–9.
7. J. D. Schulman and D. S. Karabinus, 'Scientific aspects of preconception sex selection' (2005) 10 (Suppl. 1) *Reproductive Biomedicine Online* 111–15.
8. V. A. McKusick, *Mendelian Inheritance in Man: A Catalog of Human Genes and Genetic Disorders*, 12th edn (Baltimore: Johns Hopkins University Press, 1998).
9. H. L. A. Hart, *Law, Liberty, and Morality* (Stanford, CA: Stanford University Press 1963); R. Taylor, *Freedom, Anarchy, and the Law* (Totowa, New Jersey: Prentice-Hall, 1973); T. C. Grey, *The Legal Enforcement of Morality* (New York: Knopf, 1983); D. Lyons, *Ethics and the Rule of Law* (Cambridge: Cambridge University Press, 1984); J. Feinberg, *The Moral Limits of the Criminal Law. Volume One: Harm to Others* (New York: Oxford University Press, 1984); R. A. Epstein, *Principles for a Free Society. Reconciling Individual Liberty with the Common Good* (Reading: Perseus, 1998).
10. E. Dahl, 'The presumption in favour of liberty. A comment on the HFEA's public consultation on sex selection' (2004) 8 *Reproductive Biomedicine Online* 266–7.
11. E. Dahl, 'Procreative liberty: the case for preconception sex selection' (2003) 7 *Reproductive Biomedicine Online* 380–4.
12. The Ethics Committee of the American Society for Reproductive Medicine, 'Preconception gender selection for nonmedical reasons' (2004) 82 (Suppl. 1) *Fertility and Sterility* 232–5.
13. A. Sen, 'Missing women – revisited' (2003) 327 *British Medical Journal* 1297–8.
14. E. Dahl, 'Preconception gender selection: a threat to the natural sex ratio' (2005) 10 (Suppl. 1) *Reproductive Biomedicine Online* 116–18.
15. E. Dahl, M. Beutel, B. Brosig and K. D. Hinsch, 'Preconception sex selection for non-medical reasons: a representative survey from Germany' (2003) 18 *Human Reproduction* 2231–4; E. Dahl, K. D. Hinsch, M. Beutel and B. Brosig 'Preconception sex selection for non-medical reasons: a representative survey from the United Kingdom' (2003) 18 *Human Reproduction* 2238–9; E. Dahl, R. S. Gupta, M. Beutel *et al.*, 'Preconception sex selection: demand and preferences in the United States' (2006) 85 *Fertility and Sterility* 468–473.

16. E. Dahl, K. D. Hinsch, B. Brosig and M. Beutel 'Attitudes towards preconception sex selection: a representative survey from Germany' (2004) 9 *Reproductive Biomedicine Online* 600–3.

17. Human Fertilisation and Embryology Authority, *Sex Selection: Options for Regulation* (London: Stationery Office, 2003).

18. Genetics & Public Policy Center, *Preimplantation Genetic Diagnosis: A Discussion of Challenges, Concerns, and Preliminary Policy Options Related to the Genetic Testing of Human Embryos* (Washington: Johns Hopkins University Press, 2004).

19. E. Dahl, B. Brosig, H. R. Tinneberg and S. Grüssner, 'Gender preferences and demand for sex selection: a survey among pregnant women in Germany' *Human Reproduction* (in press).

20. R. Steinbacher and F. D. Gilroy, 'Preference for sex of child among primiparous women' (1985) 119 *The Journal of Psychology* 141–7.

21. H. Statham, J. Green, C. Snowdon and M. France-Dawson, 'Choice of baby's sex' (1993) 341 *Lancet* 564–5.

22. C. Hank and H. P. Kohler, 'Gender preferences for children in Europe: empirical results from 17 countries' (2000) 2 *Demographic Research* 1–21.

23. P. Liu and G. A. Rose, 'Social aspects of 800 couples coming forward for gender selection of their children' (1995) 10 *Human Reproduction* 968–71.

24. M. A. Khatamee, A. Leinberger-Sica, P. Matos and A. C. Weseley, 'Sex selection in New York City: who chooses which sex and why?' (1989) 34 *International Journal of Fertility* 353–4.

25. F. J. Beermink, W. P. Dmowski and R. J. Ericsson, 'Sex preselection through albumin separation of sperm' (1993) 59 *Fertility and Sterility* 382–6.

26. E. F. Fugger, S. H. Black, K. Keyvanfar and J. D. Schulman 'Births of normal daughters after MicroSort sperm separation and intrauterine insemination, in-vitro fertilization, or intracytoplasmic sperm injection' (1998) 13 *Human Reproduction* 2367–70.

27. G. Pennings, 'Family balancing as a morally acceptable application of sex selection' (1996) 11 *Human Reproduction* 2339–45.

28. P. Jha, R. Kumar, P. Vasa *et al.*, 'Low male-to-female sex ratio of children born in India: national survey of 1.1 million households' (2006) 367(9506) *Lancet* 211–18.

29. S. S. Sheth, 'Missing female births in India' (2006) 367(9506) *Lancet* 185–6.

30. Kusum, 'The use of pre-natal diagnostic techniques for sex selection: the Indian scene' (1993) 7 *Bioethics* 149–65.

31. G. Benagiano and P. Bianchi, 'Sex preselection: an aid to couples or a threat to humanity?' (1999) 14 *Human Reproduction* 868–70; J. Quintavalle, *Sex Selection: Choice and Responsibility in Human Reproduction. A Response of Comment on Reproductive Ethics (CORE) to the Human Fertilisation and Embryology Authority's (HFEA) Public Consultation on Sex Selection* (London: CORE, 2003); V. Hudson and A. Den Boer, *Bare Branches: The Security Implications of Asia's Surplus Male Population* (Cambridge: MIT Press, 2004).

32. E. Dahl, 'No country is an island: a comment on the House of Commons' report Human Reproductive Technologies and the Law' (2005) 11 *Reproductive Biomedicine Online* 10–11.

33. M. Darnowsky, 'Revisiting sex selection: the growing popularity of new sex selection methods revives an old debate' (2004) 17 *GeneWatch* 2–11.

34. N. P. Chico, *Confronting the Dilemmas of Reproductive Choice: The Process of Sex Preselection*, Accession Number AAG8926403, University of California, San Francisco, 1989.

35. Frank J. Sulloway, *Born to Rebel: Birth Order, Family Dynamics, and Creative Lives* (New York: Pantheon Books, 1996).

36. T. Powledge, 'Unnatural selection: on choosing children's sex', see note 3 above, 198.

37. J. Savulescu and E. Dahl, 'Sex selection and preimplantation diagnosis: a response to the Ethics Committee of the American Society for Reproductive Medicine' (2000) 15 *Human Reproduction* 1879–80.

38. For further reading on the ethics of sex selection I recommend M. A. Warren, *Gendercide: The Implications of Sex Selection* (Towota, New Jersey: Rowman & Allanheld, 1985); P. Singer and D. Wells, *The Reproduction Revolution: New Ways of Making Babies* (New York: Oxford University Press, 1985); J. Savulescu, 'Sex selection: the case for' (1999) 171 *Medical Journal of Australia* 373–5; D. McCarthy, 'Why sex selection should be legal' (2001) 27 *Journal of Medical Ethics* 302–7; and J. A. Robertson, 'Preconception gender selection' (2001) 1 *American Journal of Bioethics* 2–9.

4

Cloning to avoid genetic disease

Lynn Gillam

The use of prenatal diagnosis (PND) and pre-implantation genetic diagnosis (PGD) is familiar as methods that a couple might use to avoid the birth of a child with a genetic condition. However, these methods will not deal with all instances of genetic disease. Both work on the basis that only *some* of the embryos formed from the gametes of the couple will have the genetic condition. In PND, the idea is that an affected pregnancy can be terminated, so that the couple can try again, in the hope that the next pregnancy will not be affected. In PGD, the idea is that some embryos will not be affected, and they will be chosen for transfer, while the affected ones will be discarded. But there are a small number of situations where this process will not work. These are situations in which, because of the precise nature of the genetic mutation involved, *all* embryos created from the gametes of the couple will inevitably have the genetic condition which the couple wish to avoid in their children.

One possible way to get around this problem would be to use reproductive cloning to create an embryo solely from the genome of the unaffected partner, thus producing an unaffected child. The advantage of this approach over the alternative approach of using donor gametes is that a third party is not introduced into the equation. The resulting child would be genetically related to the one parent, and to no-one else. There is (apparently) no 'foreign' DNA involved, there is no gamete donor with a strong genetic connection who would complicate the issue of parenthood, and no worry about the 'genetic bewilderment' of a child who has been raised by non-genetically related parents.

In this paper, I will discuss two particular situations where cloning might be considered as a preferable alternative to the standard sorting technologies of PND and PGD – avoidance of genetic disease caused by mitochondrial DNA (mDNA), and avoidance of carrier status for recessive genetic conditions.

The Sorting Society: The Ethics of Genetic Screening and Therapy, ed. Loane Skene and Janna Thompson. Published by Cambridge University Press. © Cambridge University Press 2008.

My aim is to consider whether the ethos of 'sorting' by other measures could extend as far as the acceptance of reproductive cloning to achieve such sorting in these limited circumstances. There are opposing tendencies in the literature and popular discourse that make this an interesting question. On the one hand, there is considerable support for PGD and PND, even though this involves the intended death of embryos and fetuses which are selected against. On the other hand, there is extremely widespread opposition to reproductive cloning, even though it does not (necessarily) involve the intended death of any entity. I will consider the arguments usually deployed against reproductive cloning, and show that many of them do not apply in these cases of cloning used to avoid genetic disease. However, I will argue that there are still residual ethical concerns.

The question, then, is whether these residual concerns are strong enough to make cloning ethically unjustified in these circumstances. I will argue that in the end, much depends not only on how serious the problems that I have noted are deemed to be, but also on the moral weight that is given to the two reasons for using cloning in this situation. As I indicated earlier, the use of cloning in this situation actually has two purposes: to avoid the genetic condition and also to avoid using donor gametes. The moral worthiness of both of these aims is usually taken for granted, but here I will open both up for scrutiny, on the basis that there may be reasons to question both. In general terms, if these two reasons have moral weight, then it will be reasonable to take some moral risks to avoid both genetic conditions and foreign DNA in one's children. If they do not have much moral weight, then such risk-taking will generally not be justified.

The two situations: using cloning to avoid mDNA conditions and carrier status

mDNA conditions

There are a small number of genetic conditions caused by defects in the mitochondrial DNA. This is located in every cell, in the mitochondrion, which is a structure outside the cell nucleus. It is present in the same form in every egg which a woman produces. So a woman with a mitochondrial genetic condition will necessarily pass this condition on to all her children. This cannot be avoided by PGD or PND. The reason for wanting to avoid having a child with an mDNA condition is quite straightforward (at least in common sense terms). These are nasty conditions, which tend to be very debilitating from an early age, are often difficult to treat and shorten the child's life. Making sure that one's child would not have such a condition seems to be a morally worthy aim; although of course, due to the now-familiar non-identity problem, it cannot be construed in terms of avoiding harm to the child that is to be born, since that child is a different child from the one who would have been born with the mDNA condition, and now will not be born at all. (I will say more about the non-identity problem later).

In these cases, one possible option is use of donor eggs fertilized by the male partner's sperm. But that has the disadvantage in the eyes of many couples that it introduces an outsider's DNA into the family they are trying to create (and it would have to be a real outsider – the woman's mother, sisters and maternal cousins would have the same mDNA defect, and would not be suitable egg donors). Using donated gametes has a number of documented problems,[1] including the desire of some donors for an on-going relationship with the child, difficult decisions about whether or not to tell the child his or her true genetic origins, and fear of effects of both telling and not telling on the psychological wellbeing of the child (such as so-called genetic bewilderment), and the health of relationships in the family. In order to avoid this, and also avoid the genetic condition, another option would appear to be reproductive cloning.[2] A clone of the male partner would not have the disease, and would also not have any foreign DNA. The process of cloning (as the technology currently stands) would require an enucleated egg, so a third party would still be needed, but foreign DNA could still be avoided. The nuclear DNA, which forms the vast bulk of the DNA in a cell, is removed when the nucleus is removed, and if the egg donor were a close female relative of the male partner (such as his sister), her mDNA would be identical to his, since it is passed on in its entirety from a mother to all her children. In this situation, it would truly be the case that no foreign DNA would be involved.

If a sister or a maternal female cousin is not available as an egg donor, and an unrelated donor has to be used, then foreign DNA would not be entirely eliminated, as a tiny proportion of the DNA would come from the donor. However, arguably the genetic connection would be seen as small and unimportant by the parties involved. Most people are probably unaware of the mDNA, and in any case, it does not play a role in determining the visible characteristics that people tend to look for in their genetic children. Nevertheless, the egg donor, although having contributed only a tiny amount of DNA, has taken considerable effort, undergoing the pain of egg retrieval and the risks of superovulatory drugs, to donate an egg. She may feel some claim or connection to the resulting child, out of proportion to the genetic connection alone. In this case, the aim of avoiding having a third party with some grounds for a claim over the child would not be met. Despite this, the couple may feel that this is sufficiently close to what they intended, that it would serve their purposes, and decide to go ahead. And in any case, this may be a problem only associated with the early stages of cloning. With constant advances in technology, it may turn out that an enucleated egg is not required for cloning, and a somatic cell from the male partner could be made to do the job. If this were to happen, then no third party would be required at all.

Carrier status

A number of genetic diseases, the so-called 'recessive' genetic diseases, occur when an individual has two copies of a faulty gene. Cystic fibrosis (CF) is one well-known example of a recessive condition. A child born with CF has inherited one copy of the faulty gene from his or her father and one from his or her mother, who both have a single copy of the faulty gene. They are termed 'carriers'. In high-school biology, it is said that carriers are not affected at all by the single faulty gene, but this is not in fact correct. The single faulty copy does in most cases have an effect, and it may be a positive rather than a negative one. A single copy of the CF gene, for example, is protective against diarrhoea, a major cause of infant mortality. The benefit to carriers is a probable reason why the carrier rate of some genetic conditions is so high, relatively speaking, in that it gives carriers a survival advantage, and hence a reproductive advantage. Carriers are generally unaware of their carrier status, unless they have had a genetic test, or have given birth to a child with a recessive genetic condition.[3] Carrier testing is usually prompted by the diagnosis of a genetic condition in a new baby in the family. The physical effects of being a carrier, positive or negative, are not significant enough to be noticeable to the individual.

Although carrier status for CF, or other recessive conditions, is not a health problem as such, parents may still wish to avoid having children who are carriers. A person affected by CF has two copies of the gene, and will necessarily pass on a defective gene to all of his or her children. They will all be carriers. Now that people with CF are living much longer, thanks to earlier diagnosis, better therapy, better nutrition and more treatment options (such as lung transplantation), it is becoming more common for them to consider having children. There are also more and better options for dealing with the diminished fertility that affects many people with CF. Hence the question of what sort of children to have when one has CF is beginning to arise. People with CF or other recessive genetic conditions may want their children not to be carriers for two different types of reasons. One is concern about the psychological and social effects on the child of being a carrier, and the other is the existential concern of the affected parent about removing the disease from their family. In the CF situation, parents would presumably want their children to know they were carriers, so that when they come to have children of their own, they can take steps to avoid having a child affected by CF. This could be done by asking a prospective partner to have a test to find out his or her own CF carrier status, and then either avoiding having children with this partner, or making sure to use PND or PGD. But even if children of carriers were not told of their carrier status by their parents, they would be very likely to find out anyway. Rudimentary genetic knowledge that the children would acquire at school would be enough for them to work out that they are carriers.

The psychological effects of *knowing* that one is a carrier can be negative for some people (for example, feeling at risk, feeling 'tainted', feeling unhealthy), though others are psychologically unaffected. And even though there are ways for carriers to make sure their own children do not have CF, this is still not a pleasant process. For example, in a relationship, a carrier has to make a decision about when to disclose his or her carrier status and how to have the unavoidable discussion about children, PGD, PND and abortion. (This also makes unplanned pregnancies even more stressful, if the discussion has not occurred). Parents may want to make sure their children do not have these experiences. Parents may also have their own concerns about passing on defective genes, even if they don't cause disease directly. The parent who has CF may want to make sure it 'ends with me' (a response that is not uncommon with adults diagnosed with Huntington disease, and deciding not to have children). The affected person may want to wipe the condition off the face of the earth, or at least out of his or her family, so that his or her children never have to face the *possibility* of having a child with CF.

Using cloning to avoid carrier status would work in much the same way as for avoiding mDNA conditions. Take, for example, a couple where one has CF and one doesn't, and they want to have children. Every child born to that couple would be a carrier, since every child would inherit one copy of the CF gene from the parent with CF. As with the mDNA situation, this could be avoided by cloning the unaffected partner. If the unaffected partner were male, his wife's eggs could be used, since CF genes are carried in the cell nucleus, not the mitochondria, and the nucleus would be removed during the cloning procedure. This would mean that absolutely no foreign DNA would enter into the equation. If the unaffected partner were the woman, she could use her own eggs, and the resulting child would be a full genetic copy of her, mDNA included. Hence, in the case of avoiding carrier status, the aim of avoiding foreign DNA in the child would be fully achievable, as would the aim of avoiding a gamete donor with possible claims over the child.

Objections to reproductive cloning and their applicability here

It can be seen that, in both these scenarios, there are plausible reasons for a couple to want to use reproductive cloning. But there are a number of well-known objections to reproductive cloning, so the first task must be to see whether these objections are applicable to the scenarios we are considering, and if so, what is their force.

Objections to cloning can be divided into three main groups: (i) inherent objections to the very nature of cloning; and consequentialist-style concerns about (ii) the effects on the child and (iii) the possible effects on society and future generations more broadly. I will consider each of these in turn. There is another set of objections, relatively under-emphasized in the literature, which relates to risks to those involved in the process; for example, harm to egg donors from superovulatory drugs used to ensure a maximum number of eggs for harvest, and harm to women to whom embryos are transferred, assuming that multiple embryo transfers would be required and that perhaps a number of terminations for fetal abnormality would be done. I set this group of objections aside because they are risks that competent adults have agreed to undertake, and it would be unjustifiably paternalistic to prevent them from taking risks for what they perceive to be a worthwhile goal. (It is, of course, morally required to ensure that they are fully informed of the risks, and make a free choice.)

Inherent wrongness of cloning

It is sometimes argued that cloning is unnatural or constitutes playing God.[4] These vague objections have also been raised against other forms of reproductive technologies, and have been shown elsewhere[5] to be flawed. It is highly problematic to produce an account of what constitutes 'unnatural-ness' or 'playing God', without including very many practices that we regard as totally morally acceptable (such as the whole practice of modern medicine). Trying to exclude the morally acceptable things from the definitions results in indefensible arbitrariness. Even if the definitional problem could be solved, there is the difficulty of giving a coherent and theoretically plausible explanation of why 'playing God' or doing something 'unnatural' is a moral wrong at all. So I will not consider these types of objections further.

A more specific objection to cloning per se is that it violates the autonomy of clones,[6] or violates their right to genetic uniqueness.[7] These are also poorly founded objections. Genetic uniqueness is not required either psychologically for the development of personal identity, nor morally as the basis for moral status, as the case of genetically identical twins shows. Identical twins are perfectly capable of developing and exercising their own autonomy, and there has never been any attempt to argue that they do not have a right to this autonomy, just because they are twins. It is hard to even spell out what a right to genetic uniqueness might entail, let alone on what moral value it might be based, given that autonomy is not threatened by lack of genetic uniqueness.

All of these objections would apply to reproductive cloning to avoid genetic conditions, but in my view they are poor arguments, and do not provide any plausible moral objection. So I will not consider them further.

Negative effects on the child

More frequently, objections to reproductive cloning are based on negative effects on the cloned child. These are better arguments, and I will discuss them in more detail, but it should be noted that they require a notion of non-personal harm (or wrong-doing not based on causing harm), such as that advanced by Parfit.[8] This is because any child born by cloning would never have been born at all without cloning, and so is not harmed by having been cloned unless his or her life is so bad as to be not worth living. If the possible ill-effects on children provide the explanation for why it is wrong to create cloned children, it must be on the principle that it is wrong to create children who have less good lives than other children whom one might have created, rather than on the principle that it is wrong to harm children, since the cloned children have not themselves been harmed. (Using Parfit's idea, Savulescu has coined the term 'Procreative Beneficence' to refer to the idea that people have an obligation to have the best children they can.[9] I will use this term later, in evaluating the moral weight of reasons for cloning to avoid genetic conditions.)

The 'effect on children' arguments fall into three groups: physical effects, psychological effects and moral wrong.[10] Physical effects may be caused in a number of ways. The process of cloning is not likely to be error-free. Lots of embryos are likely to be created before a viable one is produced (in the case of Dolly the sheep, it was hundreds[11]). Some chromosomal abnormalities could be detected by PGD before implantation, and the affected embryos discarded. But given the lack of knowledge and control over the process of embryonic and fetal development, there may well be physical problems which only become apparent later. If they are detectable by ultrasound during pregnancy, a decision could be made to terminate the pregnancy and try again. Whether this loss of embryonic and fetal life counts as morally serious will depend on one's view of the moral status of embryos and fetuses. But regardless of this, there will also be a significant probability of ill-effects to the child after birth. These could include severely shortened lifespan, if the telomere theory turns out to be correct,[12] degenerative neuromotor conditions that start in infancy, high rates of cancers and so on.[13] The results of attempts at animal cloning give some indication of what the problems might be, but not certainty, since the outcomes are so variable across different species.

The physical ill-effects argument obviously does apply to cloning to avoid genetic conditions, since the actual process of cloning is the same no matter what the purpose of it, and hence must be considered. The ill-effects have to be weighed against the possible advantages of having been cloned rather than conceived some other way. An attempt to do this will be made below, once psychological ill-effects of cloning have been considered.

The possibilities for psychological harms of reproductive cloning are nicely captured in Søren Holm's concept of 'a life in the shadows',[14] which is his term for being raised as a copy of someone else. This postulated harm is based on presumed motives of the parents and their likely attitude and behaviour towards the cloned child. A cloned child who is intended as a copy of someone else would presumably be brought up with this as the prime goal. He or she will not have the opportunity to chart his or her own course in life, follow his or her own interests and so on. These concerns are perhaps strongest when a child is cloned from an adult deliberately chosen for specific characteristics (such as great musical or mathematical or sporting talent[15]), with the idea (even if due to a misguided belief in genetic determinism) that the child will have the same desired characteristics as the 'clonant'. If parents are aware of the role of environment in personal development, this will only make things worse. The chosen interests and talents can be cultivated and environmentally enhanced from birth, rather than waiting for them to appear spontaneously, and the child's life will be even more controlled. This is likely to be frustrating and psychologically damaging to the cloned child. Similar effects, but perhaps reduced in magnitude, can be envisaged where the child is the clone of a deceased older sibling, brought into the world as a replacement. The older the sibling was at the time of death, the more the objections to being raised as a copy would apply. If the sibling was only a baby at the time of death, there is probably much less to copy, but there is still the problem that the child might feel (and indeed be) loved more as a replacement of someone else than for him- or herself.

A life in the shadows is not only psychologically damaging, but can also be seen as a moral wrong to the child,[16] in that it is contrary to the child's right to an open future.[17] This is a much more sophisticated and solid objection than the one discussed earlier, that being genetically identical to someone else in itself negates one's autonomy. Here the idea is that it is the upbringing of a child who is genetically identical to someone else that will undermine or frustrate the child's right to future autonomy. Choices would not be made for the child with a view to providing opportunities, keeping options open and setting the child up to have a wide range of choices about how to live when he or she is an adult. Instead they would be made to channel the child in a predetermined direction, narrowing down opportunities rather than opening them up.

These objections to reproductive cloning do have some validity, to the extent that parents' motives are as they are assumed to be. However, the life in the shadows argument is not really applicable to our scenarios of cloning to avoid genetic conditions, since the aim of this cloning is in fact to make sure that the child is *not* like one of the parents in a small but very significant way. The method of achieving this (whilst also avoiding foreign DNA) happens to involve making the child a genetic copy (almost) of the other parent. But this is an incidental side effect, not the purpose of the cloning. Where cloning is done to avoid genetic conditions, there would appear to be no more reason to fear a life in the shadows upbringing for a cloned child than for a non-cloned child.

In summary, then, there would undoubtedly be a risk of physical harm to children born as a result of cloning done to avoid genetic conditions. It seems much less likely that there would be psychological risks, since all the postulated psychological harms are not relevant to this type of cloning. Now to assess the moral significance of the physical risks, it is necessary to weigh these against the advantages of using cloning to avoid genetic conditions. In carrier status avoidance, the advantage to a cloned child relates mainly to not having to face difficult choices in finding a partner and having children of their own. It is hard to say how much of an advantage this would be for any particular child, since individual responses are likely to vary so much. The more direct benefit really accrues to the parent who achieves his or her aim of putting an end to the transmission of the bad gene, but this is in a different moral category – advantage to the parent does not directly justify disadvantage to the child. It is thus hard to come to any general conclusion about how the risks weigh up against the benefits for the cloned child in this situation. However, given that the benefits are mostly psychological, it could be argued that these could be brought about in other ways, whereas at this stage there is no known way of counteracting the possible physical harms of cloning. This would suggest that the benefits do not outweigh the possible harms.

In cloning to avoid having a child with mDNA disease, the beneficial effects would be both physical and psychological; the cloned child would not have a serious, debilitating and life-shortening genetic condition, and also would not be born into a situation where parenthood could be seriously contested or disputed. However, given the unknown side effects of cloning, the cloned child could well be physically worse off than a non-cloned child that the couple could have had using donor gametes, but also be physically better off than the child they could have had by conceiving naturally. The weighing up of physical harm is made difficult here because the possible harms of cloning are speculative: there is no evidence which could provide the basis for a sensible comparison between the unknown risks of cloning and the well-established ill-effects of having an mDNA condition. Probably the best that could be said is that there are serious possible harms on both sides, and it is not clear that one outweighs the other. When the psychological benefit to the cloned child of not having the emotional and social issues facing children conceived by gamete donation is added into this mix, the balance perhaps starts to tip in favour of cloning. But this is hardly definitive, given all the uncertainties involved.

Negative effects on society more broadly

Many of the postulated negative effects on society[18] of reproductive cloning work on the assumption that cloning would be a widespread practice. It is claimed that wide-scale cloning would encourage a culture of enhancement, would widen the gap between rich and poor, could be abused for evil purposes[19] (such as the creation of warriors and drones), or would narrow the gene pool and hence make the human race vulnerable to environmental change. Another, much more sound, objection is that it would be a wasteful use of resources. However, these objections do not readily apply to cloning to avoid genetic conditions, as these conditions are relatively rare, and use of cloning to avoid them would only ever be a small-scale practice.

It may still be argued that small-scale reproductive cloning to avoid genetic conditions will be the thin end of the wedge, and will inevitably lead either to wide-scale cloning, with all its alleged problems, or to other objectionable practices, such as making many clones of famous and successful individuals. However, this sort of slippery slope argument is not very convincing, as it gives no evidence to support the claim that the slide would occur, nor even any explanation of why it is a plausible speculation. Would many people want to use expensive reproductive technologies, with low success rates, involving medical procedures just to have a child who is a copy of someone else? The whole rationale for setting up IVF was to allow infertile couples to have their own genetic children, and on-going demand for IVF has confirmed that this is a commonly and strongly held objective. Why should we think that millions of people would suddenly want to use a cloning and IVF procedure to have a child that is not genetically related to themselves?

In short, concerns about the broad social effects of reproductive cloning are not only implausible, but also do not apply to the specific instance of cloning to avoid genetic conditions.

The positive case for cloning to avoid genetic conditions

I have identified some reasonable concerns about the potential ill-effects of cloning on the cloned child, both physical or psychological. But what is the moral significance of these ill-effects? They are hardly likely to be sufficient to make the child's life not worth living, and hence make cloning a clear wrong-doing to the child who is cloned. Nor is it obvious that they would breach even the less strict version of Savulescu's proposed duty of Procreative Beneficence, which requires parents to produce children who will have the best possible lives, unless they have good principle-based reasons to have children who will have less than the best, but still good, lives.[20] However, this does not show that there is no moral problem at all. The concerns about risks cannot be made to disappear entirely, as we simply do not have enough knowledge. So there is still a need to make a positive case in favour of cloning to avoid genetic conditions, to see if one can overcome the ethical problems caused by these concerns. I will go on to sketch out what this positive case might be. But first, I must deal with a possible objection to this approach.

It may be felt that looking for a positive moral reason in favour of reproductive cloning is irrelevant or misguided. The putative parents, it may be claimed, have a right to reproductive freedom, which includes the freedom to decide whose gametes and which technologies will be used to produce their children. It does not matter what their reasons are for their reproductive decisions: they must be allowed to make and act on them regardless. I am not very sympathetic to this argument, since I believe that the so-called right to reproductive freedom is exaggerated both in its scope and its moral significance. Hence I argue that the desire to use cloning to avoid genetic conditions and at the same time avoid foreign DNA is in need of some defensible basis, which would make it an ethically reasonable thing to do, in the face of moral reasons *not* to do it. This must be more than simply a claim to a reproductive right.

The positive case for cloning here has two pillars – one is the wellbeing of the child to be born, and the other is the parents' desire not to use donor gametes. On the basis of Savulescu's Procreative Beneficence, it would seem that concern for the wellbeing of the cloned child is morally relevant (and not just a morally neutral whim of the parents). The non-identity problem should not lead us to conclude that wellbeing is irrelevant except in the case where the life of the cloned child is not worth living. However, on the same basis, for the 'wellbeing of the child' argument to be a compelling reason in favour of cloning, it would have to be the case that the wellbeing of the cloned child would be greater than that of the non-cloned child that the couple might have had instead. As I indicated above, it is not obviously true that this would be the case. We simply do not know enough about human cloning at this stage to be able make any sort of definitive judgement on this matter.

The moral weight of the other pillar of the case is even less convincing, I believe. It is certainly true, in Western countries at least, that people want children who are genetically related to them, and will go to great lengths (in emotional and financial terms) in using IVF and other types of reproductive technology to have them. The option of adoption is not favoured (although the shortage of children for adoption, the difficulty of being accepted as a candidate for adoption, and the bad press that adoption has received may be strong contributors to this). But this does not show that the desire to have genetically related children is morally significant, such that it is right to accept moral losses to fulfil it. Of course it is a 'natural' desire, and its existence can probably be accounted for in evolutionary terms (the genes most likely to be transmitted to the next generation are the genes of those who prefer their own genetically related child). But the same can be said of the desire to rape – this is natural and confers an evolutionary advantage, as it maximizes the number of offspring a male can produce. But no-one concludes from this that men should be assisted to fulfil their desire to rape. More is needed than an appeal to naturalness.

Another possible argument is that children's welfare is greater if they are raised by their genetic parents. As a conceptual claim, this just does not make sense. Why should it be that genetic parents are *necessarily* the best parents? As an empirical claim, it is also relatively weak. Whilst there is a wealth of studies coming to a range of different conclusions on this matter, there is overall little evidence to support the claim that, in practice, genetic parents do a better job than non-genetic parents.[21] Indeed, there is ample evidence from the sorry annals of child protection cases that genetic parents are not always best at protecting and promoting their child's welfare.

In contrast to the desire to have a child with a good quality of life, which has a clear and strong moral basis, the desire to have a genetically related child looks more like a morally neutral whim. As such, it is perfectly appropriate to allow people to act on it when there are no moral costs (which is, arguably, most of the time), but not necessarily when there are moral costs. Cloning does have morally relevant costs in terms of the physical and psychological burdens that it may place on the cloned child. These may be unlikely, difficult to predict and of variable magnitude, but they cannot be totally dismissed. When there are other options for having children, which do not carry these risks, it is hard to see how reproductive cloning can be justified, even for the purposes of avoiding genetic conditions in children.

Conclusion

Taking steps to avoid giving birth to children with genetic conditions is generally regarded as a good thing to do. It is true that there are some ethical concerns about prenatal screening and diagnosis, which focus on two main issues – the extent to which women are well-informed and make a free choice about using prenatal screening and diagnosis, and the implications of these practices for people with disabilities. However, the mainstream view is that women are acting quite properly, even responsibly, by taking up these technologies. If there is general social acceptance of the goal of avoiding having children affected by genetic and congenital conditions, and sound ethical arguments in favour of it, this raises the interesting question of whether it would be justified to use reproductive cloning as a method to achieve this goal. This is the question which I have investigated here. My conclusion is that using cloning for this purpose places too much importance on having children who are genetically related to the parents and not to a donor. As much as people might desire this quality in their children, their desire is not of sufficient ethical importance to outweigh the moral costs of reproductive cloning. So cloning is not an ethically acceptable way of avoiding genetic conditions, not because it is cloning per se, but because the supposed benefits of it in terms of avoiding the use of foreign DNA are not sufficiently morally weighty. There are other ways of avoiding genetic conditions which do not have the same moral costs.

NOTES AND REFERENCES

1. See, for example, A. Braeyways, 'Review: parent-child relationships and child development in donor insemination families' (2001) 7(1) *Human Reproduction Update* 38–46.
2. Note that there is no reliable evidence that reproductive human cloning has ever been attempted, let alone successfully, though there are some claimants. In 2001, Dr Severino Antinori and American reproductive physiologist Dr Panayiotis Zavos announced their intention to begin cloning humans, claiming that 200 couples had volunteered to participate in their experiment; but they have since stopped working together, and no birth of a cloned baby has been announced. The organization Clonaid, run by a religious group called the Raelians, claims to have produced a number of cloned babies, but these claims have never been verified and are almost certainly not true.
3. Carriers of thalassaemia, another recessive single gene disorder, can be detected by the much more common haemoglobin test, since they are anaemic. This blood test is routinely done in pregnancy, and for many other purposes. Hence, people can find out that they are carriers of thalassaemia without intentionally having a test for this purpose.
4. These are usefully summarized by F. Bayliss, 'Human cloning: three mistakes and an alternative' (2002) 27(3) *Journal of Medicine and Philosophy* 310–37.
5. See, for example, L. Silver, 'Cloning, ethics and religion' (1998) 7 *Cambridge Quarterly of Healthcare Ethics* 168–72.
6. R. Williamson, 'Human reproductive cloning is unethical because it undermines autonomy: commentary on Savulescu' (1999) 25(2) *Journal of Medical Ethics* 96–7.
7. See, for example, K. Evers and K. Evers, 'The identity of clones' (1999) 24(1) *Journal of Medicine and Philosophy* 67–76, and D. W. Brock, 'Human cloning and our sense of self' (2002) 296 *Science* 314–16.
8. D. Parfit, *Reasons and Persons*. (Oxford: Clarendon Press, 1984) Chapter 16.
9. See Savulescu Chapters; J. Savulescu, 'Procreative Beneficence: why we should select the best children' (2001) 13(5–6) *Bioethics* 413–26.
10. See, for example, J. Burely and J. Harris, 'Human cloning and child welfare' (1999) 25(2) *Journal of Medical Ethics* 110–12. The authors survey a variety of postulated harms to cloned children, but conclude that these do not make a persuasive case against cloning, essentially because of the non-identity problem.
11. I. Wilmut, 'Dolly: the age of biological control' in J. Burley (ed.) *The Genetic Revolution and Human Rights* (New York: Oxford University Press, 1999) 20.
12. F. Allhoff, 'Telomeres and the ethics of human cloning' (2004) 4(2) *The American Journal of Bioethics* W29–W31.
13. S. M. Rhind, J. E. Taylor, P. A. De Sousa *et al.*, 'Human cloning: can it be made safe?' (2003) 4(11) *Nature Reviews Genetics* 855–64.

14. S. Holm, 'A life in shadows: one reason why we should not clone humans' (1998) 7 *Cambridge Quarterly of Healthcare Ethics* 160–2.

15. These characteristics may be not fully or even partly genetically determined, of course, but once parents have proceeded to cloning in order to copy someone, whether themselves or another person, this consideration is not relevant. They are acting on the belief in genetic determinism and that will shape the way they raise the child, whether genetic determinism is true or false.

16. As J. Burley and J. Harris point out in 'Human cloning and child welfare' (1999) 25(2) *Journal of Medical Ethics* 111.

17. D. Davis, 'Genetic dilemmas and the child's right to an open future' (1997) 27(2) *Hastings Center Report* 7–15.

18. For instances of these alleged effects on society, see D. Brock, 'Cloning human beings: an assessment of the ethical issues pro and con' in M. Nussbaum and C. Sunstein (eds.), *Clones and Clones* (New York: W.W. Norton and Company, 1998) 153–4.

19. See, for example, the objections listed by F. Bayliss, 'Human cloning: three mistakes and an alternative', note 4 above, 327.

20. J. Savulescu, 'Procreative Beneficence: why we should select the best children', note 9 above, 424.

21. A. J. Cherlin, 'Going to extremes: family structures, children's wellbeing and social science' (1999) 36(4) *Demography* 421–8.

5

Procreative Beneficence: reasons to not have disabled children

Julian Savulescu

The reasonable man adapts himself to the world; the unreasonable one persists in trying to adapt the world to himself. Therefore all progress depends on the unreasonable man.[1]

Introduction

Couples (or single reproducers) have a moral obligation to strive to have disability-free children. Disability is sometimes argued to be a social construction. It is said that disability should be removed by altering social institutions and circumstances. At least in some circumstances, I argue that biopsychosocial correction of disability is needed, where disability is removed by altering our biology or psychology, or selecting our children.[2]

The principle of Procreative Beneficence

In a previous paper, I sketched what I called the principle of Procreative Beneficence:

couples (or single reproducers) should select the child, of the possible children they could have, who is expected to have the best life, or at least as good a life as the others, based on the relevant, available information.[3]

This principle is novel in one way. It claims that we have a good reason to select which child we have. Many people deny this. They claim we should not select our children. According to folk morality, we should accept whichever child Nature or God gives us as a gift. There is a significant distance between the principle of Procreative Beneficence and folk morality. Can Procreative Beneficence be defended?

The Sorting Society: The Ethics of Genetic Screening and Therapy, ed. Loane Skene and Janna Thompson. Published by Cambridge University Press. © Cambridge University Press 2008.

There are many objections to this principle. I will address these. But let me first clarify this principle and give an expanded formulation.

The nature of the obligation to have the best child

We should do what we have most reason to do. This obligation is not some deontological obligation but rather a rational obligation. If A and B are identical in all regards except one, and A is superior in that regard to B, we have a very strong reason to choose A. Economists call this Pareto optimality. A is Pareto optimal to B. If we have two embryos which in all respects appear the same, except B has a state which is a disability, then we have a strong reason to choose A. We should choose A. If I am looking at two television sets which are in all regards identical, the same price, same brand and model and so on, except that B has a dent in the side while A does not, I have a reason to choose A. I should choose A. Of course, A might not work, but I have no reason to believe that A is any more likely to not work than B before I buy it.

By 'should' in 'should choose', I mean 'have good reason to'. I will understand morality to require us to do what we have most reason to do. Some people deny this. They claim that there are other reasons besides moral reasons, such as prudential reasons. For these people, my arguments establish what we have most reason to do, what we should do, all things considered. '*In the absence of some other reason for action*, a person who has good reason to have the best child is morally required to have the best child.'[4] (Italics added.)

The idea is a simple one. There is a normative force pushing us to choose the direction of selecting the best child like a vector going, say, east. But there are other normative considerations which also have normative forces which push us in other directions. Whether we finally go east or how far east depends on the number, direction and force of these other moral vectors. But there is some normative force pushing east. In philosophical terms, the claim that parents ought to have the best child means there is a *pro tanto* reason to have the best child or 'other things being equal', they should have the best child.

What might these other considerations or normative reasons be? There are many: welfare of the parents, welfare of others, harm to others, direct and indirect. As rational animals, we should respond to the weight of reasons.

Expected value

The principle of Procreative Beneficence states that we should select the child who is expected to have the best life. This principle is mistaken if 'expected to have' is interpreted as 'will have'. We cannot know which child will have the best life. Those born with the greatest gifts and talents may squander them while those born to great biological and social hardship may overcome enormous obstacles to lead the best of lives. It is indeed hubris to believe that we, with our limited knowledge and power (unlike God), could ever know who will have the best life. However, we should do what we can to give our children the best life they can have.

This approach derives from decision-theoretic consequentialism. The standard way of making decisions under uncertainty is to choose that option which maximizes expected value. While this may not be the way we make decisions in ordinary life, it is one standard norm of rationality for how an ideal agent who has no computational limitations should make decisions. In general terms, consequentialism instructs the agent to list all the relevant possible courses of action and the possible outcomes of each action. (This strictly includes *all* possible outcomes or consequences that stem from this action, no matter how far in the future.) The agent must then estimate the probability that each outcome of each action will occur, given that the action in question is taken, and assign values to each possible outcome, calculating the expected value of each possible outcome. This is the product of the value of that outcome and the probability of it eventuating, given that a particular action is taken. The agent should then calculate the expected value of each action (this is the sum of expected values of each of the possible outcomes (or consequences) of that action) and then choose the action with the greatest expected value. In the case of selection and reproductive decision-making, one important outcome of interest is how well a person's whole life goes, that is, wellbeing.

We use this approach often in our decisions in a rough and ready way. Consider a person trying to decide whether to have a knee replacement for arthritis. The decision will usually be made by weighing the pros and cons, how extreme these are and how likely they are. One needs to know how bad the pain and disability currently are, how much they will be alleviated by the operation, how likely the operation is to be successful, what the risks of the operation are and how bad the complications might be, how much the operation costs, in money and time, and the consequences of this, and what the costs or benefits of waiting are. In this way, a person can answer the question: is it better to have a knee replacement or not?

This approach can be formalized. The great golfer Tiger Woods is reputed to have had laser surgery to give him better than 20/20 vision. Imagine someone like Woods, a professional golfer wanting to win the British Open, but who is also knowledgeable about decision theory. He is trying to decide whether to have laser surgery to give 20/20 vision. The following figures are purely hypothetical.

Assume that without surgery, his life will go very well and he will win many golf tournaments. If 1 is the perfect life, his life overall will be of value 0.96. If he has laser surgery, he will win slightly more tournaments. His life will be slightly better 0.97. However, there is a risk (1:1000) that the surgery will damage his eyesight and he will win slightly fewer tournaments and his life will go slightly less well (0.95). The expected value of life without surgery is 0.96.

The expected value of life with surgery is 0.96998. Even though the benefits of surgery are small, it is rational to have the surgery given its risks are very small. As the probability of harm rises, or it becomes more serious, there is less reason to opt for surgery.

In this way we can decide whether to have an intervention – if its expected value in promoting our welfare is greater than the alternatives. Sometimes the results of such calculations, in their rough and ready form, are surprising. Some people request the amputation of a healthy limb because they believe its removal will make their lives go better. They believe that amputation has a greater expected value than remaining with four healthy limbs. Examples are people who request amputation to function as beggars, or those who feel more complete with an amputation or as a means to sexual gratification.[5] Missing a limb is a disorder but these people believe that having a disorder has a greater expected value for their lives than having four normal limbs. These people are not necessarily irrational. If one is destitute, and can make a sufficiently large amount of money out of the sale of a kidney, the expected value of life without a kidney may be overall greater than life with two kidneys. Whether a person has most reason to give up a limb, have a sex change or sell a kidney depends on the circumstances. But in some cases, these decisions will be rational. (In others, not.)

Disability and expected value

Consider Sharon Duchesneau and Candy McCullough, the deaf lesbian couple who sought to deliberately create a deaf child by using sperm from a deaf male donor. I have discussed this case previously, arguing that there is not good reason to select a deaf child but that since the child would not be harmed, the couple should be at liberty to conceive a deaf child. They act wrongly but permissibly.[6] There is a vigorous debate around whether deafness is a disability or a difference or variant of normal. In part, this is a terminological dispute which turns on how disability is defined. What matters is not whether deafness is a disability but whether it is bad and should be avoided if possible. The answer to this question turns, in significant part, on the expected value of a life with deafness compared to hearing life. One way to answer this question is to ask: should a deaf person attempt to have his or her hearing restored? This is like the question: should a normally sighted person attempt to achieve better than 20/20 vision?

The answer depends on many variables including: the value of hearing life, the chance of the operation giving hearing, its risks, the value of a deaf life and the risks and benefits of any other courses of action. The value of a hearing life (like a life with better than 20/20 vision) depends on how that hearing enables one to realize various possible good lives, and the probabilities of achieving these.

Let's call something a **disability** if it will reduce the goodness (value) of a life (disability in the intrinsic sense) and/or reduces the chances of a person realizing a possible good life (disability in the instrumental sense).

Is deafness a disability in this sense? Deafness might be both intrinsically and instrumentally bad. It reduces the goodness of a life by preventing that person access to the world of sound. A deaf person cannot hear music, the sound of wind, the crack of thunder or the seductive whisper of a lover. The human voice is a fundamental part of the human condition and verbal communication a characteristic part of human culture. If these are intrinsic goods, deafness is intrinsically bad.

But more importantly, deafness also reduces the chances of realizing a good life in many other ways because it makes it harder to live, to achieve one's goals, to engage with others in a world which is based on the spoken word. It is harder to get a job, harder to move in the world, harder to respond to emergencies where the alarm is aural, and so on. Sometimes people in the Deaf community deny these claims. They claim that deafness represents a unique culture that can only be fostered by being deaf. They claim that deafness does not reduce the value of life or necessarily make it more difficult to interact with others because signing is a unique form of communication which does not require hearing and which creates its own world of advantages. Even if signing is an effective and unique form of communication, the claims are mistaken. Hearing children of deaf people can learn to sign and communicate with their deaf parents, just as children of English parents can learn Chinese as well as English. It is surely better to have the capacity to speak two languages rather than one, to understand two cultures rather than one. It would be disabling for a child of English parents living in China if the child only spoke English, even though it might be easier for her parents to communicate with her. If signing does have the benefits which deaf people claim that it does, hearing children and people can learn it as 'a second language' and experience its benefits as well as the benefits of being able to hear sounds, voices and music.

Disability is context dependent

It is important to recognize that disability is *context* or *environment* dependent. What makes leading a good life harder in one circumstance, may make it easier in another. For example, being outspoken may make for political success in a liberal democracy but lead to death in a totalitarian regime. The atopic tendency that leads to asthma in the developed world protects against worm infestations in the undeveloped world. Deafness would be a positive advantage in an environment of extremely loud and distracting noise.

So the definition of disability should be extended: X is a disability *in circumstances c if*: X reduces the goodness of a life and/or X reduces the chances of a person realizing a possible good life.

In order to judge which conditions constitute a disability, we need to predict what the likely context or environment is likely to be. Is it better to have brain or brawn? On average, it is likely that having brains will be of greater advantage in our technologically advanced society than brawn. The answer might be different in other communities at other times. Brains might be not much use when all you have to do is plough a large field by sunset.

Importantly, people with disabilities can have very good lives. There are many other goods in life besides the goods which hearing sound realizes and deafness, although making life in our world more difficult, does not make it impossible. Just as being very poor or having very little education or having a chronic disease make it more difficult to lead a very good life, so too do other disabilities. But they do not remove opportunity altogether unless very extreme. There are countless examples of people with disabilities who have had very good lives and who have achieved much. Well-known examples are Beethoven, Toulouse-Lautrec and Stephen Hawking.

Disability and capabilities

According to Amartya Sen, our wellbeing is a function of our capabilities and functionings. A capability is the capacity or potential to be able to do something or experience something. A functioning is the actual doing or experiencing. For example, being able to hear is a capacity. Hearing beautiful music is a functioning. On my view, our wellbeing solely consists in our functionings. Capabilities have only instrumental value. Consider the following argument.

Jane is one year old. Jane's parents are given the choice to enhance her musical abilities and her ability to do maths. They choose to enhance both abilities because they believe she will be better off having the capacity to be good at music and at maths. Jane never develops her talent to do maths but she becomes an outstanding concert pianist, composing many of her own works.

At the end of her life, we would judge how Jane's life went on the basis of her functioning and flourishing as a musician (and her other flourishings). Her having an unrealized talent for maths does not imply her life went better.[7] *Ex ante*, capacities matter. *Ex post*, they do not. The reason for this is that capacities increase opportunity and the chance of a better life, a life with more wellbeing, but they do not constitute it.

Disabilities can have both intrinsic disvalue and instrumental disvalue. Being deaf is bad in itself because necessarily one cannot experience the world of sound. It is also instrumentally disvaluable, making it harder to have relationships and achieve things in the world as it is. Some capabilities are so necessarily tied with wellbeing in our world that we might be tempted to accord them instrinsic value: being able to hear means that one will hear many beautiful things whatever one's life. The capacity will, to some extent, be necessarily realized.

The instrumental disvalue of disability is the same as the instrumental value of capacity. Disability (in its instrumental role) or capacity do not guarantee a worse or better life; nor do they constitute a worse or better life. They are probabilistically related to achieving a good life. That is why people seek to avoid disability and admire capabilities and potential.

Disability and capability are mirror images of each other. Disability is the feature of a state which makes it less likely that we will achieve a good life. Capability is the feature of a state which makes it more likely that we will achieve a good life. Disability is the glass half empty; capability, the glass half full.

Disability is ubiquitous: all of us are disabled and fail to lead the best life

Procreative Beneficence faces another related objection. One might argue that there is no such thing as a better or best life. This, as I have argued, is false. A life without being able to hear, or see, is deficient in an important way. Another example of a disability is asthma, a condition which I have. Asthma makes breathing more difficult in certain environments commonly encountered in the developed world – dusty or smoky environments, places with pets. It makes it more difficult to enjoy the company of others, achieve one's goals, do physical activity if it is difficult to breathe (or hear or see). In this way, *all of us are disabled in some ways* which make it more difficult to lead a very good life.

Another objection is that it is impossible to achieve the best life. This is virtually always the case because, amongst other things, we lack complete information and the ability to process such information. But it is a feature of all decision-making in a less than ideal world. We are never sure that we have performed the act which has the best consequences, or bought the best house or the best television set or helped our friends as much as we could, even if we wanted to. This is an objection to act consequentialism and to any theory which aims to bring about a certain state of affairs in a probabilistic world. It applies whenever we try to do our best or merely try to affect the world. We can rarely if ever do our best, but we can try. We cannot be certain of the effects of our actions – we can only rationally estimate them.

This is a version of a familiar objection to consequentialism and there are many responses. In relation to selecting children, we should aim to limit our choices to selecting the best child bearing in mind the opportunity cost of this action, including the opportunity cost of gathering more information. For example, we should not sacrifice everything in pursuit of having the best child, but act in a way that is reasonable. This is the same as a person who seeks to do a good turn for his or her friend. He or she should not give up everything for the sake of his or her friend, but do only what is reasonable. Our obligations to have the best child are limited by our other obligations and the costs to us and to others of pursuing that end. There must be limits on what harms parents can be expected to bear to have a child with a better prospect of a very good life, just as there are limits on what parents must sacrifice for the welfare of their children, or friends for each other. Indeed, just as there are constraints and limits on the pursuit of the best lives for ourselves.

The opportunity costs of maximization provide a reason to submaximize. But when there is no consideration of opportunity costs, and we could costlessly maximize, there is only one reason to submaximize: it relates to our present pattern of cares and concerns which orient the direction of our lives.

Harms and opportunity costs

Melo-Martin claims that Procreative Beneficence is flawed because it will require otherwise fertile women to have IVF and pre-implantation genetic diagnosis, and this will be expensive (if not out of reach of most people in the world) and harm women.[8] But when we understand that Procreative Beneficence is a species of decision-theoretic consequentialism, we can see that this claim is mistaken. One must consider the harms as well as the benefits to all parties concerned. And one must consider the opportunity costs of pursuing the best child.

If the harm to women of pursuing the best child is significant, they should not seek to have the best child. But when a couple is already undertaking fertility treatment, they should use all available information about their embryos to select the best one, if such a procedure does not add extra harms to them or their embryo.

Another formulation of the principle of Procreative Beneficence would be:

couples (or single reproducers) should select the child, of the possible children they could have, who is expected to be as free from disabilities (as defined above) as possible, based on the relevant, available information, subject to the costs of pursuing that goal.

Disease is a disability

It is important to recall that we all suffer from disabilities as I have defined them, i.e. conditions inherent to our nature (biological, psychological or other) that either reduce the value of our lives or that make it more difficult to realize (in the sense that they reduce the chances that we will achieve) a good life. Asthma, a lame foot, pigheadedness and weakness of will are all disabilities on this definition. Indeed, on this account, disease is bad and should be avoided for instrumental reasons – it reduces the chances of us achieving the best life. Disease is, generally, a disability.

What constitutes a good life is a deep philosophical question. According to hedonistic theories, what is good is having pleasant experiences and being happy. According to desire fulfilment theories, and economics, what matters is having our preferences satisfied. According to objective theories, certain activities are good for people – developing deep personal relationships, developing talents, understanding oneself and the world, gaining knowledge, being a part of a family and so on. We need not decide which of these theories is correct in order to understand what is bad about ill health. Disease is important because it causes pain; it is not what we want and it stops us engaging in those activities that give meaning to life. Sometimes people trade health for wellbeing – mountain climbers take risks to achieve their goals, smokers sometimes believe that the pleasures of smoking outweigh the risks. Life is about managing risk to health and life and promoting wellbeing.

Such an approach allows us to explain why we treat disease and the extent to which we believe a diseased person should be treated. The extent of the claim that a diseased person has to be treated depends on the extent to which that disease is a disability. Some diseases have so little impact on a person's life that such diseased people have very little claim to treatment. A symptomless disease which had no impact on the length or quality of a person's life would be irrelevant.

However, this is not always the case. Take colour blindness. This is generally seen as a very mild disease because it has little impact on a person's life. In some cases, it can be a sign of more serious generalized disease. Although colour-blind people experience the world differently, generally they are able to function normally and are able to discern relevant colours when it matters. Although it is a disease, it constitutes a relatively mild disability. Perhaps such people cannot appreciate great paintings and visually colourful scenes in the same way as the rest of us, but it does not detract significantly from their lives. But now imagine a master painter, expert in colour who becomes colour blind. Such a person might be prepared to spend vast sums of money to correct his or her colour vision. This represents the value of colour vision to that particular person in his or her context. For such a person, colour blindness might be a severe disability.

Biopsychosocial correction of disability

Our biology evolves slowly, over thousands if not millions of years. Our social life has radically changed over the last one hundred years. Doctors are keen to tell us that our biology is not suited to our current high fat, low fibre diet and sedentary lifestyle. But our biology and psychology are probably more globally out of synchrony with our way of life. It is not merely that we are prone to 'lifestyle disease', it is that we are prone, to some degree, to lifestyle unhappiness.

Whenever there is a mismatch between biology, psychology and social environment resulting in a bad life we have a choice. We can alter our biology, our psychology or our environment. This is occurring in medical practice when doctors advise diets which are low in fat, high in fibre, high in antioxidants, which lower our cholesterol and which basically mimic the diet which our bodies are designed to tolerate. The most extreme example of this is the stone age diet, which attempts to replicate the diet of primitive man. But another approach is not to change our environment (in this case diet) but to change our biology through drugs. The 'polypill' is designed to allow our bodies to tolerate a modern diet and lifestyle by chemically lowering our cholesterol, our blood pressure, etc.[9]

When it comes to having children, we can attempt to alter our environment to suit our individual children or we can select children to suit our environment. Some people have attempted to change their environment. For example, since asthma develops from an immune response which was originally beneficial in protecting us against worm infestations, one doctor is attempting to replicate this condition in the developed world by introducing benign worms into the intestines of asthmatics. But another solution would be to have children without the predisposition to asthma.

My own view is that all routes must be considered. In some cases, it is reasonable and practicable to alter the environment. For example, it is preferable to give women fair and equal opportunities than to resort to sex selection. But in other cases, it is going to be difficult to change the modern environment to allow all possible people to flourish. For example, in areas of high ozone layer damage, it may be more effective to choose children with increased melanin pigment in their skin to protect them from the sun, rather than attempting to close the hole in the ozone layer in that area or enforcing the use of sunscreen, coverage of the skin and fear of the sun.

Because disability is context-dependent, we must ask what the context is likely to be or could be. Deafness is likely to remain a disability in the likely possible worlds. It would be wonderful if every child learnt to sign and all modes of verbal communication were accompanied by appropriate non-verbal communication for the deaf. But resource limitations simply prevent us bringing about such a world.

The consequences of being deaf in our world can be tragic. On the night of 10 April, 2003, a school for deaf and mute children in Makhachkala in Russia caught fire. Twenty-eight children aged 7 to 14 died and more than 100 were injured. 'Several children, some naked, jumped through windows to escape the inferno.' Rescuing the children was hampered because 'each child had to be awakened individually and told in sign language what to do'.[10] No doubt some of these children would have lived if they had not been deaf.

When is disability 'socially constructed'?

All disability is in one sense socially constructed and recognizing this often leads to an argument that society should be changed to enable the disability to be overcome. As I have argued earlier, when a condition constitutes an impediment to wellbeing (that is a disability), we have a choice: to change biology, psychology or society. But when should we change society rather than psychology or biology? When it comes to existing people, there is an argument in favour of changing society because changing biology or psychology in such cases would change the nature of the identity of existing individuals. To correct a defect and cause a person to cease to exist and another one to come into existence who is better, is like killing one and replacing that person with another person. It would be no cure of my headaches to offer me a brain transplant, even if the (different) transplanted individual never suffered headaches, because I would no longer exist.

What of correcting intellectual disability? For example, imagine that we could remove the extra chromosome 21 from every cell of an adult with Down syndrome, restoring normal intellectual capacity. Call this procedure 'chromosome resection'. Would chromosome resection be a desirable cure for Down syndrome? It is reasonable to assume that this procedure would bring about sufficient psychological alteration to constitute an identity-altering intervention. If one questions this, ask: what would I be like if I had Down syndrome? Such a person might be so thoroughly different, it would not be me – it is hard to imagine what my character would be like if I suddenly developed Down syndrome. Most of the things I care about, I would no longer care about. My goals and life plans would go unfulfilled. Whatever the value of the life, it would not be mine.

If chromosome resection is identity-altering, or at least sufficiently biography-altering, should we cure Down syndrome? Here the answer turns on whether the intellectual disability afflicting a human being with Down syndrome is so severe as to render that human being a non-person. In cases of very severe disability, where the afflicted individual has little awareness of self or others, chromosome resection would constitute killing a non-person and replacing him or her with a person. In cases of minor intellectual disability, chromosome resection would do little to alter identity. In moderate cases, it would alter identity – it would constitute killing a person and replacing him or her with a different person. While the replacement of one person with a different person is morally questionable, it is less clear whether replacement of non-persons with persons is wrong. The upshot of this argument is that cure of mild and severe intellectual disability may be permissible, but correction of moderate intellectual disability is more problematic.[11]

Besides being identity-altering, there can be other reasons against biological or psychological interventions, such that social interventions are more favourable. For example, they may be riskier or less likely to succeed. In such cases, we should consider social interventions, which may be safer, more effective and more just (based on the limitations of resources and the likelihood that the benefits will outweigh the harms).

Consider the argument that mixed race couples have a moral obligation to have children with lighter skin because having darker skin socially disadvantages people. There are good reasons to prefer social correction of any disability that race represents. Skin colour does not confer any intrinsic disadvantage in the way that deafness or paralysis prevents a person engaging with the world. Skin colour is an instrumental disability. Moreover, there are many reasons to think that the social institutions that make race a disadvantage should be changed for existing people. The costs of removing unjust race-based social discrimination are small when compared with the costs of altering society to give the deaf, or blind, or paralysed equal opportunities. These latter states constitute both intrinsic and instrumental disabilities. Altering the world so that paralysis is no longer a disability is expensive and bound to be only partly successful. Distributive justice, the fair use of our limited resources, provides a reason against social correction of the disability that deafness and paralysis cause. It is more cost-effective to biologically correct these disabilities, that is to cure them, rather than changing society to accommodate them.

Moreover, in some places, race or skin pigmentation can have biological value. In the Northern Hemisphere where there is limited sunshine, fair skin provides an advantage in allowing the skin to produce more vitamin D necessary for calcium metabolism and strong bones. Near the equator, where the sun is strong, fair skin is a disadvantage because such people are more predisposed to skin cancer. In conditions of strong sunlight or ozone depletion, dark skin is a biological advantage. Having fair or dark skin can be a capability or a disability, depending on the natural and social environment.

It may not be possible to make social changes that will give everyone a good life. In reality it will take years for social institutions to change enough to reduce unjust discrimination, and there is then a reason to choose children who will have better lives within the unjust system. If state schools are not going to improve, there is a reason to send one's child to a superior private school even if one believes that justice requires better state schools. In some cases, the disabled are right – we should change society. But in other cases, we should change ourselves or change our reproductive practices in order to have children who are better suited to the likely social environment in which they will live.

Treatment vs. enhancement

According to regulatory authorities such as the Human Fertilisation and Embryology Authority in the UK and the Victorian Infertility Treatment Authority in Australia, selection against serious diseases is permissible, but selection for non-medical characteristics such as sex, height or intelligence is not. That position reflects the common belief that there is a morally significant difference between medical treatment and enhancement. Such a distinction is difficult to defend. Why do we allow selection for serious diseases but not for minor disorders, such as cleft lip, if they are both diseases/disorders? Moreover, as I have argued, the explanation grounding the moral imperative for the treatment of diseases is not that they are diseases, but that they have a significant effect on length and quality of life, that is, how well our lives go. But if that is the reason for allowing selection, anything which can affect how well a person's life goes is up for selection or manipulation.

Many people are attracted to the basic parenting principle that parents should give their child the best possible start in life (given constraints on the cost to them or others of pursuing such a goal). Procreative Beneficence reflects this principle to some degree: that we should choose children who start off life with the best chance of having the best life. Isn't the goal of being a good parent to raise a child who ends up having a good life?

Some might respond that this is not the goal. The goal is to provide the circumstances for this particular child to realize his or her full potential and to have the best possible life, for him or her. If this were the principle of good parenting, it would imply there is nothing wrong with selecting in favour of a child who will be born with a serious genetic disease, provided one does the most one can to alleviate that child's suffering. Some hold this extreme view, but most people can see the rationale behind selection against disease genes. It is a small step from this to recognizing that there is nothing special about disease and that all selection to reduce disabilities, however minor or subjective, is permissible.

The strongest argument: Pareto optimality and opportunity restriction

I was always thought of as an unusual character. I just thought I was rational. In my final medical exams, I wore industrial strength ear muffs. I wanted to be deaf for the duration of the examinations. And I was. I was hearing, but I chose to be deaf, for a reason and for a time.

While I can be deaf when I want to be, deaf people cannot be hearing when they want to hear, or not at least to the same extent. Cochlear implants and hearing-aids are not as effective at providing hearing as ear muffs and polyurethane ear plugs are at removing noise.

Let us assume that the nature of good life is obscure (though I strongly doubt this). Let us assume, for argument's sake, that we do not know whether it is better to be deaf or hearing. If this is the case, it is better for individuals to choose for themselves whether it is better to be deaf or hearing. Since the hearing can become deaf, but the deaf cannot become hearing (to the same extent), it is better that our children hear. If they really believe that it is better to be deaf later in life, they can wear ear muffs or have the nerves to their ears cut.

We should distinguish between opportunity-restricting/enhancing states and opportunity-altering states. Deafness is absolutely opportunity-restricting in the world as it is likely to be and so is a disadvantage. Being able to hear is a Pareto optimal state with respect to capacities. If one can hear, one can become deaf. But if one is deaf, one cannot becoming hearing to the same extent. We need to consider the costs and reasons of biological, psychological and social modification. But in some cases, certain biological states provide greater opportunities,[12] and biological intervention is preferable.

If hearing-aids were perfect (like say optical correction for short or long sight), then being deaf would not be opportunity-restricting. But they are not perfect. To take another example, some people believe they are better off with only three limbs rather than four. I seriously doubt whether it is a better life to have three limbs. But let us grant their claim. It is still better that our children be born with four limbs – they can always choose to have a limb amputated later in life. But we are not like salamanders who can regrow limbs. Prosthetic limbs are not the same as real limbs. Being limb-less is opportunity-restricting in a way that being limb-ful is not.

Many states will not be real disabilities because they are fully alterable. I recall one friend who had entirely normal sight but started wearing glasses with no optical correction simply to look more intelligent and professional. She could enjoy the benefits of wearing glasses without short sight. Optical correction of refractive errors is now so effective that they are close to not being disabilities at all. It is true that one cannot be a fighter pilot without 20/20 vision, but with such exceptions, refractive errors rarely interfere with a person's life. Once laser eye surgery is perfected, they will not be disabilities at all. Another example is breast size. Breast size is significantly genetically determined. But we do not have a good reason to choose the breast size of our children as there are such effective methods of making breasts larger or smaller. A person can choose their own.

In many cases, undeveloped talents and potentials are opportunity-restricting – by simply not developing the talent or potential, one loses an opportunity that would otherwise have been available. But if you lack a talent or potential, there is nothing you can do about it; you are stuck in that state whether you like it or not. Capacities that can remain unrealized or be ablated are autonomy-enhancing. Far from frustrating any 'right to an open future'[13], such selection facilitates a more open future,[14] provided that they can be detected later in life.

Not all abilities and disabilities will be opportunity-enhancing or-restricting. Some are opportunity-altering – giving rise to a different set of capacities and opportunities. For example, I doubt whether having a certain height, having better memory or perfect pitch are reversible. So one must make a choice between a life of hearing pitch perfectly, and being able to detect one's own mistakes and the mistakes of others exquisitely, or a life oblivious or less attuned to these imperfections. Those states will be associated with different sets of opportunities. With our children, we may have to make a choice between different lives since people cannot move from one state to another easily or completely. Once you have better memory, it would be difficult to lose it. There are arguments to support selecting some opportunity-altering states, based on their relationship to wellbeing and the development of full autonomy (as I will argue) but they are different and arguably weaker than the arguments in favour of opportunity-enhancing states since the latter are compatible with later autonomous choice while the former are not.

The biological and psychological basis of autonomy

So far, I have offered mainly welfarist arguments in favour of Procreative Beneficence arguing that we should choose those traits which make it more likely that our children will have the best lives, in effect, minimizing their disabilities. In the previous section, I raised a different style of argument: that certain traits will increase the choices and opportunities of our children because they are opportunity-enhancing. They will increase the openness of that child's future. Potentials are important examples in this category. I wish to close this elucidation of the arguments for Procreative Beneficence with a third argument in favour of selection: the biological and psychological bases of personal autonomy.

There are certain social arrangements which are necessary to make autonomous action possible. But there are also biological and psychological preconditions for autonomy to be possible. Autonomy is self-rule or self-determination. In order to be autonomous, one requires a sense of self. This requires memory and a coherent self-concept. Autonomous choices are not mere choices – they are made according to matching the consequences of various courses of actions against one's own concept of the good life. This requires the mental abilities to form a concept of the good life and the ability to understand, remember and deliberate on the basis of relevant information. This requires certain cognitive abilities which are present in varying degrees in humans.

I cannot give a full account of the biological and psychological preconditions for being autonomous. That is another paper. But in so far as such preconditions can be elucidated, we have powerful reasons to seek them in our children. Certain forms of biological selection can be employed to increase the autonomy of our children. There are strong arguments based on the value of being autonomous for making such selections. There are genetic and other biological bases of being an autonomous individual. Just as we should provide the material bases of self-respect, we should provide the biological basis of autonomous action.

Conclusion

In this chapter, I have shown the error of Shaw's statement with which I started: 'The reasonable man adapts himself to the world; the unreasonable one persists in trying to adapt the world to himself. Therefore all progress depends on the unreasonable man.' Those claims are wrong. Sometimes it is rational to adapt, biologically or psychologically, to the world. Sometimes it is rational to change the world. And sometimes, we should accept things just as they are. Which course of action or inaction we choose depends on the benefits and risks, the opportunity costs and the context. What we must do is consider all options and make an active choice which reason supports. In particular, we do have good reason to select the best children, those with the least disabilities. We should no longer leave reproduction thoughtlessly to chance.

NOTES AND REFERENCES

1. G. B. Shaw, *Man and Superman* (Cambridge, Mass: The University Press, 1903; Bartleby.com, 1999) at Maxims for revolutionists no.124, www.bartleby.com/157/6. html (accessed 29 February 2008).
2. This paper draws on material to be published in J. Savulescu and G. Kahane, *Procreative Beneficence and Disability, The Moral Obligation to Have Children with the Best Chance of the Best Life* (under review) and J. Savulescu, 'In defence of Procreative Beneficence' (2007) 33 *Journal of Medical Ethics* 284–8. The account of disability is further developed and modified in G. Kahane and J. Savulescu, 'Welfarist account of disability' in K. Brownlee and A. Cureton (eds.) *Disability and Disadvantage* (Oxford: Oxford University Press, forthcoming).

3. J. Savulescu, 'Procreative beneficence: why we should select the best children' (2001) 15 *Bioethics* 413–26.
4. *Ibid.*
5. C. Dyer, 'Surgeon amputated healthy legs' (2000) 320(7231) *British Medical Journal* 332.
6. J. Savulescu, 'Education and debate: Deaf lesbians, 'designer disability,' and the future of medicine' (2002) 325(7367) *British Medical Journal* 771–3.
7. Although paradoxically, a life of unrealized talent seems to go worse than a life which never had that talent. This only speaks to the central claim – that wellbeing is what matters. The former person could have had a better life, with more wellbeing, while the latter could not. The former was worse off than he or she could otherwise have been; the latter was not.
8. I. de Melo-Martin, 'On our obligation to select the best children: a reply to Savulescu' (2004) 18(1) *Bioethics* 72–83.
9. N. J. Wald and M. R. Law, 'A strategy to reduce cardiovascular disease by more than 80%' (2003) 326 *British Medical Journal* 1419–23.
10. AFP, 'School fire kills 28', *Herald Sun* (Melbourne), 11 April 2003, 38.
11. The situation is even more complex. When intellectual disability is so severe as to render an individual's life not worth living, there would be a moral obligation to cure that disability even if it were identity-altering.
12. Jonathan Glover considers people who have become sighted after being blind in order to evaluate the state of blindness. This is a different epistemological argument about the value of these states, as judged by people who have been in both (J. Glover, *Choosing Children* (Oxford: Oxford University Press, 2006) 53–4).
13. J. Feinberg, 'The child's right to an open future' in W. Aiken and H. LaFollette (eds.), *Whose Child? Parental Rights, Parental Authority and State Power* (Totowa, New Jersey: Rowman and Littlefield, 1980), 124–53. D. Davis, 'The right to an open future' (1997) 27(2) *Hastings Center Report* 7–15.
14. There is a problem when it comes to choosing the capacities or states of children, as I have argued elsewhere (R. Robertson and J. Savulescu, 'Is there a case in favour of predictive testing of children?'(2001) 15 *Bioethics* 26–49). We cannot undo the effects of childhood development. Thus a hearing adult cannot choose to have been a deaf child, and to have accommodated to deafness from birth. So being hearing is less than Pareto optimal.

6

Reprogenetic technologies: balancing parental procreative autonomy and social equity and justice

Leslie Cannold

Introduction

In-vitro fertilization techniques and pre-implantation genetic diagnosis (PGD) currently allow parents to select an embryo for implantation in a woman's womb to avoid the resulting offspring suffering from, or being a carrier of, an inherited genetic disorder. In the future, it may be possible to screen and select embryos for non-disease genetic traits or predispositions. As well, some moral philosophers anticipate technologies that will enable parents to create embryos with a reduced susceptibility to disease as well as with valued non-disease characteristics such as height, intelligence, heterosexuality, impulse control, resistance to alcoholism, maternal behaviour, extraversion and athleticism, to name a few. In this chapter, I will refer to all such reproductive and genetic selection techniques as 'reprogenetic technologies'.

If science does deliver the knowledge necessary for parents to create embryos free from disease and with desired non-disease traits and predispositions,[1] questions arise as to whether parents would be morally obligated to use such technologies or, if no such obligations exist, whether it might still be morally permissible for them to choose to use them. Even if obligations can be founded or parental choice justified, we must ask whether society has good reason to encourage or allow untrammelled parents to use the technologies or whether some restrictions or even prohibitions may be justified.

To explore these questions, I will examine two claims. Firstly, the assertion that as long as such technologies are safe, parents are morally obligated to use them to create a child who, in the words of one reprogenetic advocate, has the 'best opportunity of the best life'.[2] Secondly, the weaker contention that while not obligated to use such technologies, parental choice to do so is morally

The Sorting Society: The Ethics of Genetic Screening and Therapy, ed. Loane Skene and Janna Thompson. Published by Cambridge University Press. © Cambridge University Press 2008.

permissible. Finally, I will consider the proper role of society in a world capable of developing reprogenetic technologies that enable parents to select or create offspring free of disease traits and with desired non-disease ones. Is society ever justified in restricting parents' freedom to use some or all of these technologies or in banning their use all together?

Those asserting a parental obligation to use reprogenetic technologies or the permissibility of parental choice to use them often put their case in terms of the value of such enhancements to resulting children. They also cite the entitlement of individuals and couples to procreative autonomy in defence of their position. They suggest, and I agree, that the fundamental importance of procreative autonomy places the burden on those who wish to constrain parental decision-making to offer substantial and pressing reasons for doing so, and to utilize the least restrictive tools available to achieve their ends.

However, despite the strength of claims from procreative autonomy, my analysis suggests there may be a number of substantial and pressing reasons for social interference in the freedom of parents to choose reprogenetic technologies. These reasons include the futility or socially negative consequences of parental decisions to create children with certain characteristics, the potential for use of the technologies to increase the disadvantages already experienced by environmentally impoverished children and the disproportionate and, in some cases, unjust burden an obligation, or even choice, to use the technologies places on women and other oppressed groups. In addition, society may be justified in interfering if the decisions some parents make to use the technologies changes the nature of the decisions facing others in ways that curtail the latter's freedom to say 'no' to use.

In the face of such concerns, I contend that parents are not obligated to use reprogenetic technologies and that society may be justified in imposing some limits on parental choice to use them. The challenge is to design a regulatory scheme that places the least possible limits on parental procreative freedom consistent with meaningfully reducing the harms that seem likely to result from unfettered parental choice.

Do parents have an obligation – or is it morally permissible for them – to use reprogenetic technologies to create the 'best' children?

Arguments in favour of parental obligation or choice regarding reprogenetic technologies include appeals to consistency, parental culpability for harm to children caused by failing to use the technologies and the overriding importance of parental procreative autonomy.[3]

Consistency

Appeals to consistency suggest that humans have always attempted to select the characteristics of their children through selection of their mates. Because reprogenetic technologies simply offer parents a more reliable means of producing offspring with the best chance of success, it is inconsistent to object to the latter but not the former. Says Savulescu:

Selective mating has been occurring in humans ever since time began. Facial asymmetry can reflect genetic disorder. Smell can tell us whether our mate will produce the child with the best resistance to disease. We compete for partners in elaborate mating games and rituals of display which (sic) sort the best matches from the worst. As products of evolution, we select our mates, both rationally and instinctively, on the basis of their genetic fitness – their ability to survive and reproduce. Our (subconscious) goal is the success of our offspring. With the tools of genetics, we can select offspring in a more reliable way.[4]

However, even within a Darwinian schema, it is by no means clear that mate selection is driven by anything resembling the conscious, rational decision-making that would characterize the use of reprogenetic technologies to select the characteristics of offspring. Indeed, says anthropologist Sarah Hrdy, mate selection is only one component of natural selection: the component she calls the 'mindless' process that drives evolution:

Darwin proposed that humans, along with every other kind of animal, evolved through a gradual, mindless, and unintentional process dubbed natural selection. Morally indifferent, natural selection culls and biases life chances with the unintended result that evolution (defined today as the change in gene frequencies over time) takes place.[5]

This analysis bolsters the logical assertion that even if humans have always selected their reproductive partners for the characteristics they believe will lead the two of them together to produce the most successful children, there is something qualitatively different about women or couples using reprogenetic technologies to directly select the characteristics of future offspring. Because what Savulescu calls selective mating and parental use of reprogenetic technologies are disanalogous, the moral obligatory/moral permissibility nature of one does not imply the moral obligatory/moral permissibility nature of the other.

It is also claimed that because parents are morally obliged to give their children every possible *environmental* advantage (i.e. education), it is inconsistent to suggest they are not similarly obligated to give they children every possible *genetic* advantage. Asks Harris:

If we could engineer enhanced intelligence and health into the embryo should we not do so? If these are legitimate aims of education could they be illegitimate as the aims of medical, as opposed to educational, science?[6]

Agar makes a similar point in his quest to justify the permissibility of parental choice to use reprogenetic technologies:

If we are permitted to produce certain traits by modifying our children's environments, then we are also permitted to produce them by modifying their genomes.[7]

While it is permissible for parents to make sacrifices in order to provide their children with the maximum number of environmental advantages, there is no generally recognized moral obligation to send one's children to private school, or to give them piano or tennis lessons. If children without such advantages fail to become concert pianists or to triumph at Wimbledon, parents are not construed as having harmed them. On this reading, while Agar may be right that it is permissible for parents to maximize their child's genetic advantages by using reprogenetic technology, Harris appears wrong to suggest a parental obligation to use the technology exists.

But does society have good reason to allow parents to choose? One potentially concerning implication of allowing parents' choice is that the aggregate consequences of such choice could profoundly alter the nature of the choice left to remaining parents. Intelligent children have relative advantages over their less intelligent peers. Suppose that so many parents choose to use the technologies to produce more intelligent children that kids with enhanced intelligence become the norm and unenhanced kids dunces. In such a world, parents who don't enhance their children may no longer be seen as making the permissible choice not to genetically alter their child, but as making the impermissible choice not to act to avoid disadvantaging their child in significant and foreseeable ways. Such parents could be accused, in other words, of violating what Feinberg called the child's right to 'an open future'.[8]

Insofar as non-use decisions are understood in this way, the choice by some parents to use the technologies could result in a situation where all parents effectively wind up with an obligation to use them. This state of affairs, where individual procreative autonomy is undermined rather than enhanced, is one liberal pluralist societies may have good reason to avoid.

Parental culpability for harm

Once technology provides us with the capacity to treat a condition or to affect an outcome over which previously we had no control, non-use metamorphoses into a decision for which we bear – and are seen to bear – responsibility.[9] Sometimes, because of counter-factual stipulations made by proponents of reprogenetic technologies that in the future we will know that reproductive technologies will be 'safe' or 'risk-free',[10] this observation appears as a claim that parents must either use the technologies, or be damned for the (presumed negative) consequences of non-use:

Once technology affords us with the power to enhance our and our children's lives, to fail to do so will be to be responsible for the consequences. To fail to treat our children's disease is to harm them . . . To fail to improve their physical, musical, psychological and other capacities is to harm them, just as it would be to harm them if we gave them a toxic substance that stunted or reduced these capacities.[11]

It is true that the availability of reprogenetic technologies presents parents with options that were previously unavailable, and responsibility for the nature and consequences of the decisions they make with regard to them. However, only if it does turn out to be the case that all the technologies will be known to be safe and effective, will it be true that their availability makes parental decisions not to use them culpable.

In fact, there seems little reason to assume that use of reprogenetic technologies will, in every instance, entail lower risks and greater benefits for the resulting child, other social groups or society as a whole. Few contemporary medical interventions are seen to offer such a favourable risk:benefit ratio to every child in every situation. Instead, parental decisions about reprogenetic technologies in the future are likely to mimic their medical decision-making in other areas today: consisting of an evaluation of the risks and benefits of use as against other available alternatives – including non-use – for all concerned.

Indeed, it seems likely that many parents will face considerable challenges in making what Harris describes as 'reasonable judgements as to the various probabilities involved'.[12] This is because in the early days, there is likely to be an absence of information about the risks and benefits of reprogenetic technologies, and this will pose difficulties for parents trying to weigh up their use against better-understood alternatives, including non-use. In such cases parents may, as Harris suggests, be right to 'err on the safe side' and – in so far as the widespread view that the 'safe side' is the inaction or preservation of the status quo – avoid using the technologies.[13]

The point here is simply that while the availability of reprogenetic technologies changes the nature of the reprogenetic choices facing parents, this observation does not – on its own – support claims that parents have an obligation to use the technologies. Indeed, in the early stages of availability, a lack of information about the costs and benefits associated with use may give parents good reason to avoid using the technologies.

Procreative autonomy

Autonomy is a concept of importance in Deontological, Consequential and Virtue ethical theories, among others. Autonomous agents are ones who freely direct and govern the course of their own life and whose capacities, beliefs and values are integral to him or her and are the source from which his or her actions spring. Autonomous agents largely determine their own ends and, as such, are recognized as having the dignity that moral agency bestows.[14]

Procreative autonomy reflects the view that choices about reproduction are essential to women's capacity to achieve social equality and central to the capacity of every individual to live what they understand to be a good life.

Reproduction is an intensely personal matter, and if, when and how we parent, and the sorts of children we bring into this world, reflect our identities and express our most fundamental values and beliefs. As the bearers and usual carers of children, procreative autonomy is a value of particular significance to women. This is why, all things being equal, it is considered wrong for societies – through act or omission – to limit the reproductive freedom of their citizens in general, and women in particular.[15] The central importance of autonomy is also reflected in contemporary medical law where patient autonomy is seen as the overriding consideration even when controversial uses of this freedom – like employing pre-implantation and prenatal diagnostic techniques to diagnose and terminate embryos/fetuses – are at issue.[16]

The compelling claims of procreative autonomy put the onus on those who wish to restrict an individual or couple's procreative autonomy to demonstrate substantial and pressing grounds for doing so. As well, they suggest that where such a case has been made, society ought to employ the least restrictive tools available to achieve stated ends. However, despite the high value society rightly places on procreative autonomy, my analysis suggests that there may be a number of substantial and pressing grounds for society to restrict choice to use the technologies, or ban them altogether.

Substantial and pressing grounds for interfering in parental procreative autonomy

A pointless race to the bottom

Some attributes that parents may wish to use reprogenetic technologies to achieve for their children offer only relative, rather than absolute, advantages. They are of benefit, in other words, only because some people have them and others don't. Height may be one such characteristic. In Australian society, the tallest accrue numerous advantages over the vertically challenged, including greater odds of marrying and of earning more money.[17] Tall athletes also have considerable advantages over shorter ones in some sports.

Yet, if all or most parents use reprogenetic technologies to have taller-than-average children (essentially the definition of 'tall'), there could be substantial costs to individuals and society. This is because average heights are likely to escalate rapidly, with the resulting 'height race' potentially leading to a population that may be too tall for its own good (back problems, for instance, may be more likely amongst the very tall) and too tall for existing facilities (ceiling heights in everything from public buildings to private houses to trams, buses and cars would need to be changed). As well, taller people are likely to be more toxic to the environment with their need for more food, larger cars and bigger houses.[18] These costs would come with no corresponding benefits because whether we all end up at the same (increased) height, or with a similar bell-curve distribution of height as we currently have but with a 'taller' mean, the relative gain parents were seeking for their children by using the technologies would not be achieved.

Such consequences suggest the wrong-headedness of obligating parents to use the technologies to create children with attributes such as 'tallness', or even allowing them choice in the matter, given the high costs to such children and/or the human race.

Of course, height is a particular example, and in some cases parents could choose attributes that – even if the same choice by enough other parents cancels out the relative advantage for their child – could be neutral or even positive in terms of their impact on the human race. For instance, it's hard to see how a 'hand–eye coordination' race, or an intelligence one, would generate a similar set of physical and environmental problems for humans. Indeed, one could imagine that as a species, humans might benefit from being hyper-coordinated or super smart. However, parents would need to be informed that while they and their offspring were shouldering the risks and burdens of using the technologies, the benefits would accrue to society as a whole.

Increasing social disadvantage

One of the most potent objections facing those who contend that parents are obligated – or should be permitted – to use reprogenetic technologies to create the best child is the claim that unfettered parental access may increase social inequality.

Sociological research over many years has consistently found that as family incomes fall, there is a corresponding increase in the risks to children's health, behaviour, learning and socialization. Poverty, in other words, is the most significant risk factor children face in their struggle to achieve in life.[19]

The genetic lottery, however, allows the occasional counter-example.[20] It ensures, in other words, that every so often a child with outstanding innate musical, artistic, athletic or intellectual talent overcomes the disadvantage of poverty and succeeds in their chosen endeavour against the odds.

But what would be the fate of environmentally disadvantaged children in a world where parents are obligated or are free to choose to use reprogenetic technologies to create the best child? One option is that only wealthy parents would be able to act on this injunction or to exercise this 'choice'. In fact, even if reprogenetic services were covered by private health care, as many as 40 million Americans would be left out in the cold. If, as some predict, the technologies are not seen as insurable expenses, significantly more parents will be unable to afford them.[21]

In either case, the consequence of a user-pays approach to access would be that the children of an economically advantaged subset of the population would not only benefit from the environmental advantages their parents' money can buy – music lessons, homes in neighbourhoods with better athletic facilities and schools – but also from a genetic edge. In such a world, the odds of a poor child deprived of both a genetic and environmental edge providing a counter-example to the enrichment = accomplishment norm would be remote.

One suggested remedy to this problem is that the State provides free tickets to the poor for a lottery in which the prize is a suite of genetic enhancement services similar to ones typically purchased by the wealthy.[22] Another is that the State subsidize reprogenetic services so everyone can afford them.[23]

Neither of these responses adequately responds to the justice concerns that generated them. Typically, lotteries generate few winners. A few poor children with a winning ticket for reprogenetic enhancement is not an adequate counter to the entrenchment of disadvantage among the poor that an untrammelled user-pays system is likely to yield. The second seems better, until we consider that in a world where everyone starts life with a winning genetic hand, the relative importance of the environment in determining life achievement will increase. That is, without the occasional equalizing impact of the genetic lottery, environmental inequalities will play an even greater role than is currently the case in determining a child's life chances. This suggests that unless societies funnelling resources into eliminating genetic inequality simultaneously attack the source of environmental inequality – a doubtful proposition – widespread parental use of reprogenetic technologies could *increase* the achievement gap between the children of the wealthy and those of the poor.

These significant justice-based concerns not only suggest that parents have no moral obligation to use the technologies, they cast doubt on claims that parental choice to use them is permissible.

Disproportionate and unjust burdens on women and other oppressed groups

If embryos are going to be selected or genetically enhanced, they must first be created in vitro. To do this, parents must undertake IVF procedures that impose considerable physical burdens on women. These include serial blood tests, ultrasound examinations and surgical procedures to retrieve oocytes.[24] As well, hyper-stimulating the ovaries to produce multiple eggs increases women's risk of certain types of hormone-dependent cancers, while the costs of ovarian hyper-stimulation syndrome (OHSS) in its severe form may cause renal impairment, liver dysfunction, thromboembolic phenomena, shock and even death.[25] These medical facts suggest that there will be costs to women of complying with a moral obligation to use the technologies that do not apply to men.

It is estimated that two in every five pregnancies worldwide are unplanned though not all of these will be unwanted.[26] A moral obligation for parents to use reprogenetic technologies to enhance their children's lives would present women who discover themselves pregnant with an unplanned, but not unwanted, child with a conflict between their moral obligation to any child of theirs to use reprogenetic technologies prior to implantation to create the best child, and their obligations to the particular embryo/fetus already conceived. Another way of describing this conflict is as one between women's obligation to any child of theirs to use reprogenetic technologies to create the best child and their moral beliefs about abortion.

For Harris this potential conflict can be solved by 'people' – by which he clearly means women – avoiding unplanned pregnancy in the first place by being more 'circumspect' and less 'rebellious':

There is a deeply ingrained rebelliousness in human beings which coupled with a love of spontaneity and freshness is always likely to undermine moral imperatives that require the radical circumspection which would enable people to avail themselves of genetic protections for their children . . . Even now there is some rebellion against the moral imperatives created by our knowledge of how habits, diet, in short lifestyle, affect the course of a pregnancy. Already people resent strictures to abstain from smoking, alcohol, drugs, and other things known adversely to affect the developing human individual. Inevitably, even if genetic modification of their children is available to them, many will see the circumspection required for such protection of their children to be not only burdensome, but too great a sacrifice of other values [like] . . . reckless humanity . . .[27]

However, according to the World Health Organization (WHO), neither 'rebelliousness', a 'love of "spontaneity"' nor reckless humanity' are major causes of unplanned pregnancy. Instead, the main causes include women's/ couples' lack of access to contraception and/or information about how to use it properly, violence against women and high rates of contraceptive failure. Indeed, the WHO says that even if all contraceptive users used methods perfectly every time, there would be nearly six million accidental pregnancies annually.[28]

This suggests that no matter how deliberate, circumspect and compliant their reproductive behaviour, women will continue to experience unplanned pregnancies, not all of which will be unwanted. This means that in a world where parents are morally obligated to use reprogenetic technologies, some women will face an acute moral conflict between their parental responsibilities to any future child and their responsibilities to a particular conceived embryo/fetus.[29]

While the presence of such a conflict does not defeat claims for a moral obligation to use the technologies, it does complicate it more than advocates of such an obligation acknowledge. As de Martin-Melo notes, while the assignment of a moral obligation that disproportionately burdens women or another social group may be permissible, some justification for this unequal burden must be offered, not to mention some discussion of how women ought to resolve conflicts between competing moral responsibilities to present and future children. In the absence of such a justification and discussion, I would agree with her that the claim for a parental obligation to use reprogenetic technologies to create the best children must be 'serious[ly] question[ed]'.[30]

Women are not the only oppressed group that could be adversely affected by a parental obligation, or option, to use reprogenetic technologies. Indigenous Australians suffer disadvantage relative to the non-indigenous majority in the enjoyment of economic, social and cultural rights, particularly in the field of employment, housing, health and education.[31] Similarly, African Americans are disadvantaged relative to Whites on economic, educational, employment and other social measures.[32]

These sad facts raise the question: if science provides the tools to determine an embryo's skin colour, would dark-skinned parents (say of Aboriginal Australian or Black African extraction) living as minorities in White, racist societies be obligated to use such technologies to create a light-skinned child?

Those who argue that parents have a moral responsibility to act on relevant, available genetic information to create the child with the best chance of the best life seem committed to saying 'yes.' Yet, good arguments exist about the justice of Black parents undertaking the burdens of creating a child using reprogenetic technologies to overcome social discrimination that could be overcome by pursuing social change. Discrimination against Aboriginal Australians and Black Americans is widely acknowledged to be unjust, as well as a problem of which they are the victim, not the author. Yet, the reprogenetics 'solution' requires parents to bear the costs associated with its amelioration. These costs include fiscal costs and physical costs that, as already noted, will be disproportionately borne by Black women. As well, such parents may find the process of choosing against Black skin colour in their offspring demeaning and self-abnegating, as it arguably requires them to reject for others a genetic attribute that may be constitutive of their own sense of self and identity.

Even if use of the technologies is construed as a choice, the consequences remain troubling. If some Black parents who can afford the technology seek access to it, the likely aggregate consequences of such individual choices may – in a few generations – be a reduction in the number of Aboriginal Australians and Black Americans. It will be the children of Aboriginal and Black American parents who can't afford the technology or who choose not to use it who will appear Black with the same frequency as is currently the case. However, the use of the technologies by some Black parents will mean that children of non-users will be pursing achievement in a society in which shrinking numbers of people are Black. In this society, fewer people will be negatively affected by racism, a situation that could reasonably be expected to reduce the number of those motivated to fight for its elimination.

This suggests that use by some Black parents of reprogenetic technology to select a White child could reduce the resources available to the Black children who remain to combat the racism that led to the need for the reprogenetic solution in the first place: an outcome that could see racist incidents rise. This could cause another previously discussed problem: the creation of an effective obligation on all parents to use the technology to avoid violating their child's right to an open future.

Reprogenetic technologies: can access be both free and fair?

As I have tried to show, obligating or even allowing parents to choose to use reprogenetic technologies to have the 'best' children can result in a range of surprising, harmful and unjust consequences. These include undermining the freedom of parents to choose not to use the technologies, burdening the environment, women and other minorities and further eroding the life chances of the poor. The substantial and pressing nature of such consequences suggest that societies may be justified in regulating parental access to and use of the technologies.

Regulation need not be prohibitionist in intent or consequence, but can be procedural, enabling rather than restrictive, and can operate in ways that represent widespread moral agreement.[33] An ideal regulatory approach would be one able to maximize the procreative freedom of individuals and couples consistent with fostering the wellbeing of women, the resulting child, the human race and the environment, and ensuring equality and justice for socially disadvantaged groups.

Ensuring that women and couples are properly informed about the costs and expected benefits for all involved through the provision of professional, non-directive information and counselling services will ensure women and couples are making a substantially informed and voluntary decision to use the technologies. This could reduce the risk of parents making futile choices, or ones that damage the human race collectively, or harm the environment,

However, professional non-directive counselling is unlikely to discourage all or even most parents from making decisions to create children without socially disabling traits, because as long as it is true that able-bodied White male children enjoy an advantage in our society, parents have good reason to give their informed and voluntary consent to use of the technologies to create children with these characteristics. Similarly, in a context where only the wealthy can afford to use the technologies to create the best offspring, the relative advantages for reprogenetically enhanced children over those unable to afford a genetic edge will be real. Parents who can afford the technologies may legitimately feel that a failure to use them to create the best child will render them guilty of violating their child's right to an open future. Whether informing parents of the social inequities that could result from the aggregate consequences of their individual choices would – or even should – change their minds is unclear.

Issues of just allocation of limited resources lie at the heart of many of the concerns about unfettered parental choice about reprogenetic technologies. Is it right that individuals who suffer disadvantages flowing from unjust social conditions beyond their individual control (like racism) should fund (perhaps at the expense of their dignity and self-esteem) the cost of their children's emancipation? Is it just that societies that deliberately create the market conditions that lead to disparities of wealth stand by – or even assist – citizens to access expensive technologies expected to increase these disparities and the inequality of opportunity they produce? The only moral answer to both these questions seems to be 'no'.

What this suggests is that an appropriate regulatory response would focus as much on a right to procreative autonomy as the social and material conditions in which individuals and couples make choices. As Petchesky notes:

> [Individuals] make their own reproductive choices, but they do not make them just as they please; they do not make them under conditions they create but under conditions and constraints they, as mere individuals, are powerless to change. That individuals do not determine the social framework in which they act does not nullify their choices, or their moral capacity to make them. It only suggests that we have to focus less on 'choice' and more on how to transform the social conditions of choosing, working and reproducing.[34]

As we have seen, parents may have good reason to use reprogenetic technologies to create children with characteristics that do not always and unavoidably impede a child's opportunities for a good life, but only do so in particular and remedial social contexts i.e. Black parents choosing White children. Blackness is not an intrinsic barrier to the 'best life', but one only in societies that are racist. Societies that create barriers to achievement that give parents good reason to use reproductive technologies may be understood to have dual obligations. On the one hand, such societies are obliged to give parents – and even assist them to exercise – the freedom to respond to that society by creating offspring best placed to have the best life. On the other hand, such societies are obligated to remedy the economic, cultural, social and/or political circumstances that render particular characteristics – like Black skin – a barrier to achievement. It would seem appropriate to me that the attitudinal and concrete support societies provide for citizens to exercise reproductive choice should match – on a 1:1 basis – the attention they give to the environmental conditions unjustly constraining those choices. This means that at the same time as society is respecting the procreative autonomy of its citizens, it is working to modify the conditions that constrain their choices in unjust ways.

The randomness of the genetic lottery advantages those on the pointy end of the environmental one. Its elimination further benefits those whose wealth – because of the environmental advantages it purchases – already enjoy a lead in the competitive race for achievement. Neither making reprogenetic technologies available through a lottery nor providing universal access to the technologies through whole-scale subsidies to the poor redresses these problems.

The answer may be to require the technologies to be distributed randomly. In Australia, for instance, those with private health insurance can purchase health care superior to the public provision used largely by the poor. However, both rich and poor must queue for the crisis care provided by hospital emergency departments, which is allocated solely on the basis of medical need. This situation ensures on-going interest amongst the wealthy for the state of the public hospital system as they may be among those patients arriving in casualty wanting top quality medical care.

Perhaps capacity to pay should also be removed as a criterion for access to reprogenetic services. While need is a slippery concept in relation to these technologies, lotteries in which the entire population participate to gain access to these services would ensure their benefits and harms were fairly distributed among all members of society. At the same time, societies adopting this approach should match their dedication to remedying the impact of the genetic lottery with a commitment to ameliorating the impact of the environmental one on a 1:1 basis: a commitment the population is likely to support as long as its members are unsure when they will gain access to the promised genetic benefits offered by the technologies. Thus, at the same time as societies expend effort to ameliorate the effects of a lottery that largely damages the life chances of the privileged, equivalent efforts are made to counter the negative impact of the one that – by definition – wreaks havoc with the life chances of the underprivileged.

NOTES AND REFERENCES

1. Dr Leanda Wilton, head of Genetic and Molecular Research Laboratory at Melbourne IVF, contends that the technological know-how to test for a multiple non-disease gene – and to safely modify those genes to produce desired outcomes – is around the corner. She says that developing such a capacity will be enormously expensive and extremely time-consuming and, as far as she knows, is not currently being pursued (L. Wilton, personal communication (2004)). At a forum in Melbourne Australia in 2005 on Genetic Selection, Dr Dianna Devore of the Australian Stem Cell Centre voiced a similar view.

2. J. Savulescu, 'New breed of humans: the moral obligation to enhance' (2005) 1(1) *Ethics, Law and Moral Philosophy of Reproductive Biomedicine* 38.

3. J. Harris, *Clones, Genes and Immortality: Ethics and the Genetic Revolution* (Oxford; New York: Oxford University Press, 1998); J. Harris and S. Holm (eds.), *The Future of Human Reproduction: Ethics, Choice and Regulation. Issues in Biomedical Ethics* (Oxford: Clarendon, 1998); J. Savulescu, 'Procreative beneficence: why we should select the best children' (2001) 15(5/6) *Bioethics* 413–26; J. Savulescu, 'New breed of humans: the moral obligation to enhance', note 2 above, 36–9; N. Agar, *Liberal Eugenics: In Defence of Human Enhancement* (Oxford: Blackwell Publishing, 2004); P. Singer, *Shopping at the Genetic Supermarket* (undated) www.petersingerlinks. com/supermarketprint.htm (accessed 3 February 2004).

4. J. Savulescu, 'Genetic interventions and the ethics of enhancement' in B. Steinbock (ed.), *The Oxford Handbook on Bioethics* (Oxford: Oxford University Press, forthcoming).

5. S. B. Hrdy, *Mother Nature: National Selection and the Female of the Species* (London: Chatto & Windus, 1999) 126.

6. J. Harris, *Clones, Genes and Immortality: Ethics and the Genetic Revolution*, note 3 above, 173.

7. N. Agar, *Liberal Eugenics: In Defence of Human Enhancement*, note 3 above, 113.

8. Feinberg first posited a child's right to an open future (J. Feinberg, 'The child's right to an open future' in W. Aiken and H. L. Follette (eds.) *Whose Child? Children's Rights, Parental Authority and State Power* (Totowa, New Jersey: Rowman and Littlefield,1980)). My use of it here as a negative claim right that generates a correlative duty not to behave in any way that violates the child's capacity to have an open future. See: M. Lotz, 'Feinberg, Mills and the child's right to an open future' (2006) 37(4) *Journal of Social Philosophy* 537–51.

9. P. Lauritzen, 'Experience as truth? Feminist ethics, experience and reproductive technology' (1991) 11(1) *Bioethics News* 8–18.

10. J. Harris, *Clones, Genes and Immortality: Ethics and the Genetic Revolution*, note 6 above, 222; J. Savulescu, 'New breed of humans: the moral obligation to enhance', note 2 above, 38.

11. J. Savulescu, 'New breed of humans: the moral obligation to enhance', note 2 above, 38.

12. J. Harris, *Clones, Genes and Immortality: Ethics and the Genetic Revolution* note 6 above, 214.

13. *Ibid.*

14. R. Young, *Personal Autonomy: Beyond Negative and Positive Liberty* (London; Sydney: Croom Helm, 1986).

15. L. Cannold, *What, No Baby? Why Women are Losing the Freedom to Mother, and How They Can Get it Back* (Fremantle: Fremantle Arts Centre Press, 2005).

16. B. Bennett, 'Choosing a child's future? Reproductive decision-making and preimplantation genetic diagnosis' in J. Gunning and H. Szoke (eds.), *The Regulation of Assisted Reproductive Technology* (Hampshire, Burlington: Ashgate, 2003).

17. Every additional centimetre of height added $485AUD to an annual pay packet ('NumbersCrunch' Good Weekend, *The Age*, 7 February 2004, 12.).

18. P. Singer, *Shopping at the Genetic Supermarket*, note 3 above.
19. G. Duncan, W. Yeung, J. Brooks-Gunn, J. R. Smith, 'How much does childhood poverty affect the life chances of children' (1998) 63 (June) *American Sociological Review* 406–23.
20. I am indebted to Neil Levy for my characterization of this point.
21. M. Mehlman and J. Botkin, *Access to the Genome: The Challenge to Equality* (Georgetown: Georgetown University Press, 1998) 88.
22. *Ibid.*, Chapter 6.
23. P. Singer, *Shopping at the Genetic Supermarket*, note 3 above.
24. I. de Melo-Martin, 'Discussion. On our obligation to select the best children: a reply to Savulescu' (2004) 18(1) *Bioethics* 75.
25. *Ibid.*, 76.
26. World Health Organization, *Unsafe Abortion: Global and Regional Estimates of the Incidence of Unsafe Abortion and Associated Mortality in 2000* (Geneva: World Health Organization, 2004).
27. J. Harris, 'Rights and reproductive choice' in J. Harris and S. Holm (eds.), *The Future of Human Reproduction: Ethics, Choice and Regulation. Issues in Biomedical Ethics* (Oxford: Clarendon, 1998) 239–40.
28. World Health Organization, *Safe Abortion: Technical and Policy Guidance for Health Systems* (Geneva: World Health Organization, 2003) 12.
29. For a sustained discussion of the obligations women feel to particular conceived could-be children see L. Cannold, *What, No Baby? Why Women are Losing the Freedom to Mother, and How They Can Get it Back*, note 15 above.
30. I. de Melo-Martin, 'Discussion. On our obligation to select the best children: a reply to Savulescu', note 24 above, 72–83.
31. Aboriginal & Torres Straight Islander Social Justice Commissioner,(2002) *Social Justice Report.* www.hreoc.gov.au/social_justice/sjreport_02/chapter4.html#4.1.1.1 at 6 March 2006.
32. D. Conley, *Living in the Red: Race, Wealth and Social Policy in America* (Berkeley, Los Angeles; London: University of California Press, 1999).
33. L. Cannold and L. Gillam, 'A new consultation management process for managing divergent community views: lesbian and single women's access to artificial insemination and ARTs' in J. Gunning and H. Szoke (eds.), *The Regulation of Assisted Reproductive Technology* (Hampshire, Burlington: Ashgate, 2003) 206.
34. R. P. Petchesky, *Abortion and Woman's Choice: The State, Sexuality and Reproductive Freedom* (Boston: Northeastern University Press, 1985) 11.

7

Genetic technology and intergenerational justice

Janna Thompson

The deeds of present generations create favourable or unfavourable conditions for people not yet born. But our actions not only affect what they will experience in the future. We also make those who will have experiences, and we shape their capacities for thought and action. Genetic technology provides an ever-increasing power to determine what their nature and capacities will be. In discussions about the ethical implications of genetic technology and the restrictions that ought to be placed on research and application, philosophers have mostly worried about the impact of the technology on individuals. They have considered whether and how particular techniques would violate rights or have an adverse effect on the wellbeing or autonomy of individuals. In this chapter I will focus on the impact of genetic technology on intergenerational relationships. My aim is to determine whether these techniques, now or in the future, could create intergenerational injustices.

Generations and justice

In discussions of intergenerational justice several senses of generation come into play. In a family, those who count as members of the same generation are defined by their relation to their parents or by their position in a family tree. In a social sense, a generation is a group of individuals whose births fall within specified dates and who move through life together. How we set the parameters is an arbitrary matter. When philosophers discuss duties to future generations (and more rarely, duties to past generations) they are usually thinking of the present generation as consisting of those citizens who are now in a position to make the political decisions that can affect the interests of those who, because of their position in time, are not in a position to

The Sorting Society: The Ethics of Genetic Screening and Therapy, ed. Loane Skene and Janna Thompson. Published by Cambridge University Press. © Cambridge University Press 2008.

participate in decision-making: the very young, the unborn and the dead. The composition of the present generation is continually changing and membership of future generations is indeterminate. Nevertheless, it makes sense, at any particular time, to consider how we, as people who have the power to make decisions, should act with respect to generations who do not yet, or any longer, possess this power.

Intergenerational relationships are just when each generation does its fair share to fulfil intergenerational obligations (which can include obligations with respect to the past), maintain just institutions and practices, and ensure that members of younger and future generations obtain those things that we have reason to believe should be valued by every generation. As members of a liberal-democratic society we have a duty to do our fair share to maintain our just institutions, or to reform and change them if they are not so just, and to ensure that future citizens will be able to maintain them. Most people are also likely to agree that we have a duty to perpetuate other things of value for future generations: for example, artistic products and practices, natural environments and species and links with the historical past.

Members of a family have a duty to ensure that its younger members are properly raised and educated and that they develop the capacity to act as autonomous individuals and responsible citizens. How responsibilities for educating the young are divided between families and the rest of society, and how members of a society should share these responsibilities, are matters of contention in every liberal democracy. But political philosophers generally agree that families and other social institutions should work together to ensure that members of younger generations develop into individuals who are able to make rational decisions about their own lives, participate in their society and who have an opportunity to appreciate and enjoy the good things that it makes available.

What counts as a fair share of the intergenerational responsibility for maintaining just institutions and practices is an issue of philosophical debate. According to Rawls's 'just savings principle' each generation has an equal duty to save at a rate that takes into account its relative wealth until the society reaches a level of affluence that enables principles of justice to be applied and just institutions to be maintained.[1] But once this point is reached the duty of each generation is merely to maintain just institutions and preserve their material preconditions. It is no injustice, according to Rawls, if future generations are less well off than the present generation – so long as they are able to maintain institutions of justice.

Many philosophers disagree. Bruce Ackerman insists that in a liberal society, where no individual's interests count more than the interests of others, the next generations ought to obtain at least the same quantity of resources as present generations enjoy.[2] Robert Goodin thinks that 'considerations of intergenerational equity would demand . . . that each generation be guaranteed roughly equal benefits and that one generation may justly enjoy certain benefits only if those advantages can be sustained for subsequent generations as well.[3] Ronald Green lays it down as a basic principle 'that the lives of future people ought ideally to be "better" than our own and certainly no worse.'[4] Marcel Wissenburg requires that 'no goods shall be destroyed unless unavoidable unless they are replaced by perfectly identical goods . . .' 'No rock, animal, or plant should be destroyed, no species made extinct.'[5]

There is a tension in Rawls's account between his insistence that each generation should assume an equal burden to establish just institutions and his refusal to require that they should share an equal burden in maintaining them. In my view, we do our fair share as members of the present generation if we act to ensure that future generations will not be more heavily burdened than we are in their task of maintaining just institutions and practices, fulfilling intergenerational obligations and perpetuating those things that every generation ought to be able to enjoy. However, in what follows I will not depend on the acceptance of this account of fair shares. My aim will be to determine whether the application of genetic techniques is compatible with any reasonable conception of intergenerational justice in a liberal democratic society, including Rawls's less demanding theory.

Genetic engineering and family relationships

Much of the philosophical discussion of the impact of genetic technology on intergenerational relationships has focused on the relationship between parents and children. Though the main aim in this chapter is to consider the impact of this technology on the relationship of generations as social groups in a political society, it will be worthwhile to consider why some philosophers think that generational relationships in a family could be threatened by techniques that might allow parents to choose the characteristics of their children.

Parents, care-givers and educators have an intergenerational duty to ensure that children will be become individuals who are capable of choosing and pursuing their own good. If use of genetic technology undermines the ability or inclination of parents to ensure that their children can make autonomous decisions about how to live their lives, then it undermines in a serious way the capacity to perpetuate a liberal society. Jürgen Habermas believes that genetic enhancement techniques pose this danger by undermining the ability of children to become autonomous individuals and equal citizens.[6] His reasons are two. First of all, if a genetic programme makes a child what he or she is, then the child cannot object to, or rebel against, his or her parents' determinations, as the child can against the upbringing or schooling that they chose to provide. The child has been rendered unfit to be a free individual who can decide the course of his or her own life. And secondly, a genetic programme gives the child the status of a manufactured object – a mere extension of the will of others – and thus he or she cannot demand the respect that individuals in a liberal society are supposed to receive.[7]

Both of these objections seem to overstate the effect of genetic programming, as Habermas's critics point out.[8] A child, however programmed, will have his or her own experiences and desires. The child will develop a mind of his or her own and will have no difficulty making critical judgements about his or her parents' intentions. The child does not have to become the sort of person that they want him or her to be. The child can rebel against the parents' attempt to determine the child's future by means of genetic technology just as he or she can rebel against their attempt to determine the child's future through education. Moreover, the nature of a child's genes is only one factor influencing his or her characteristics. Genetic programming cannot make the child similar to the outcome of a manufacturing process in any way that could make him or her less worthy of respect than his or her fellows. Habermas over emphasizes the importance of genetic constitution.

However, there is a psychological truth embedded in Habermas's worries about parental influence. Parents have children often for reasons that have to do with their interests, lives and relations to others. They may hope for a child that will fulfil their expectations, continue their projects, or realize their goals. There is nothing wrong with such desires. But it is important to recognize that the desire of parents and other care-givers to exercise control over the formation of the personalities of children can be very strong. Nevertheless, parents also have an intergenerational duty to ensure that their children will develop into individuals who are capable of making rational, independent decisions about their own lives. It is not always clear when justified efforts of parents to influence the choices of their children become unjustified attempts to control their lives and limit their independence, but parents who accept liberal values must put a limit on their attempts to exercise control, particularly as their children become older.

So long as it is assumed that children will have predispositions and talents which are different from those of their parents, or different from what their parents could have anticipated, it is not so difficult to regard them as independent beings, and not so difficult to resist the temptation to be controlling. The worry about genetic enhancement that Habermas overstates is that when parents have more control over the genetic constitution of their children, the temptation to try to take control of their lives will be harder to resist. People (not just parents) will be less inclined to regard children as independent beings, they will be less inclined to fulfil their intergenerational responsibility to nurture and educate children to make their own choices about what to value – and this will be so even if they have a false idea about what genetic programming can accomplish.

A liberal society therefore has reason to be concerned about genetic enhancement projects that could encourage parents to think of children as their creations. It would be bad for the children and it might also interfere with the ability of a society to perpetuate its liberal values and institutions. There is something right about Habermas's concerns, and as we will see, his claim that genetic enhancement could undermine the very basis of a liberal society is even more cogent in the context of other intergenerational considerations.

Genetic enhancement and intergenerational justice

To perpetuate the just institutions and practices of a liberal society it is necessary to maintain the conditions in which they can exist and flourish. For example, if environmental deterioration threatens the health and economic wellbeing of members of a society, this will also threaten their ability to maintain their just institutions and the things that they have reason to value. Thus there is good reason to insist that our intergenerational duties include a duty to maintain a healthy environment. A healthy social environment is also essential for the flourishing of just institutions and the preservation of things that deserve to be valued. If crime and social unrest increase significantly, if class or race antagonism escalate, maintaining just institutions will become much more difficult – perhaps impossible. Officials will come under increasing pressure to limit liberties and to violate rights for the sake of protecting lives and property. The response of many governments to threats of terrorism – a relatively minor danger for most citizens – shows that most liberal governments are prepared to enact legislation which overrides rights that have been basic to liberal societies for generations. If the threat to a society was much greater – if there was a danger of civil disruption or if crime became rampant – it is reasonable to predict that many rights and liberties would disappear – perhaps to the extent that its institutions could no longer be regarded as just.

Genetic enhancement or genetic selection processes could make it difficult for future generations to maintain just institutions by creating problems of social order. In a society where families prefer to have male children, a programme of sex selection would result in many more male children being born than female children. In some provinces of China, for example, 120 boys are born for every 100 girls. The male children when they grow up will find it difficult to find wives and many of them will not be able to realize their desire of forming a family. The effects on society of an imbalance in the sex ratio are speculative, but it seems likely that citizens of the next generations will find it more difficult to protect the security and rights of women. Some of the gains in independence that women have achieved may be eroded. Men who cannot marry may be more inclined to engage in crime or other kinds of disruptive or anti-social activities. They may become a menace to peace and order, thus increasing the cost of maintaining institutions of justice, and perhaps threatening the very existence of these institutions. Whether these bad effects will actually occur is difficult to say. But the threat is serious enough to justify a prohibition or limitation on the use of sex selection in societies where it is likely to lead to a significant gender imbalance.

Use of genetic enhancement or selection techniques could make it difficult for future generations to maintain a just society and other things that are worthy of being valued by undermining mutual respect and trust, or the willingness of people to cooperate with each other and accept the result of democratic processes. Suppose that parents will someday be able to ensure that their children will be intelligent and highly talented – at a price. Wealthy parents will be able to afford the genetic techniques; poorer parents will not. Children from wealthy families already have many social advantages and genetic engineering will add to them. In a few generations, most of the rewarding and influential positions in the society will belong to children of the wealthy. Some intelligent and highly talented children will continue to be born in poor families, but it is likely that governments will be less inclined to ensure that these children have a good education or an equal opportunity to obtain rewarding positions. Providing good education to children who generally have inferior abilities will be regarded as a waste of public money, and since the wealthy will be in positions of power and will want to ensure that the money they spend on their children gets the desired results, it will be difficult for those concerned with justice to prevail.

So in succeeding generations we can expect that class divisions will become more pronounced. The poor will obtain few of the benefits of their society and will have no hope that things will be better for their children. This may result in social disruption: class warfare motivated by the desperation of those who lack hope for the future. But even if there is no threat of civil disturbance, we can expect that some of the prerequisites of a liberal society will be increasingly absent. The privileged 'upper class' will be predisposed to think of themselves as naturally superior to members of the 'lower class'. They may come to think of people of the lower class as a subject population that needs to be controlled. They may decide that people of the lower class are not fit to vote or hold office – or at least that their votes should count for less than the votes of intelligent, knowledgeable people of the upper class. People of the lower class are likely to find it increasingly difficult to be treated as equal participants in a common political enterprise.

What this picture of a possible future suggests is that Habermas could well be right to claim that genetic engineering has the potential to undermine the conditions on which the existence of a liberal society depends. Justice to future generations requires that this technology be controlled in a way that ensures that this result does not occur.

Prerequisites of justice

The problems discussed in the last sections arise because genetic techniques produce situations that predispose parents, citizens and governments to act unjustly: to violate rights, to fail to treat others as self-determining individuals, to adopt measures and practices that are illiberal and undemocratic. 'Enhanced' children do not lose the ability to become self-determining individuals. The worry is that parents will fail to treat them as such. Similarly the 'unenhanced' people of the lower class deserve to be treated as equal citizens; they deserve an equal opportunity and a fair share of social resources. The problem is that the class system produced by genetic engineering predisposes upper class citizens to treat them unjustly, and these upper class citizens have put themselves into a position where they have the power to enforce their will. My arguments thus depend on empirical assumptions about how people would respond to changing social circumstances.

Perhaps parents and citizens are more disposed to be just than these arguments assume. Indeed, if we suppose that they inhabit Rawls's ideal world where everyone always satisfies the requirements of justice, then parents will never unjustly control the lives of their children and members of a society will never take away the rights of others. In these ideal circumstances, genetic manipulation, it seems, would create no threat to institutions of justice and thus no intergenerational injustice.

However, suppose that the members of the upper class in the society that I have imagined feel somewhat uneasy about the injustices that they are committing and are also worried that their lower class subjects will someday revolt. They now have the genetic technology that can render the next generation of lower class people into docile, stupid beings who are content to do menial tasks and are incapable of understanding abstract notions like 'freedom' and 'equality'. They use this technology, and when other people in the world complain about the injustice they have committed they offer one or more of the following arguments.

All judgements about justice or injustice, they say, assume that certain conditions are fulfilled. Hume rightly insists that there are necessary conditions for the application of requirements of justice and they include a rough equality of capacities and aptitudes.[9] But equality does not exist, even roughly, between the capacities of upper and lower class people. Or they claim that in order to have rights of citizenship people have to be able to claim them, and then point out that lower class people are not interested in, or capable of, claiming citizenship rights. Or they simply argue that it would be as ridiculous to extend rights of citizenship to lower class people as it would to chimpanzees or young children.

What the people of this imagined society have done is to undermine in a fundamental way the prerequisites of institutions of justice. They have realized

Habermas's worst nightmare. They have made it impossible for liberal-democratic institutions of justice to be maintained. My story of the future may strike many people as extremely fanciful, perhaps even ridiculous from a scientific point of view. Nevertheless, it is interesting that some well-known works of fiction have provided accounts of future societies which have in a similar way rendered themselves incapable of maintaining liberal-democratic institutions. In H. G. Wells' *Time Machine*, upper and lower classes have become so distinct that communication and cooperation, required by any institutions of justice, are completely impossible.[10] The beastlike Morlocks can only be the predators of the refined Eloi. In Aldous Huxley's *Brave New World* a combination of genetic techniques and pharmaceuticals makes lower classes unfit for anything else but their menial function.[11] The idea that all people are created equal has no relevance to this society, and liberal democratic practices are out of the question.

These stories, whether fanciful or not, can be used to make an important point about the meaning of intergenerational justice. They do so by bringing to our attention a puzzle about the application of a theory of intergenerational justice. If we could render the next generations incapable of being subjects of justice, as we understand it, then it seems that we have done nothing unjust according to our standards. If requirements of justice do not apply to them, then these requirements cannot be used to judge that we have treated these future generations unjustly. Thus we have not violated requirements of intergenerational justice. This conclusion is counter-intuitive. It seems right to insist that we would violate requirements of justice in a fundamental way (as Habermas implies). The difficulty is to explain why this is so.

Justice and lifetime-transcending interests

In *A Theory of Justice*, Rawls establishes principles of intergenerational justice by imagining that heads of families are placed behind a veil of ignorance which keeps from them information about the situation of their family in their society. Motivated by a concern for their descendants, the contractors advocate a principle that requires each generation to save to establish, and then to maintain, institutions of justice for future generations. Concern for their descendants is sufficient to prevent the contractors from bringing about conditions which would make them unfree or incapable of appreciating things of value. However, Rawls's critics complain that by privileging the position of those who have families he violates the liberal requirement of neutrality concerning ideas of the good.[12] Not everyone in a society has, or expects to have, children to care about.

However, we can meet this criticism by introducing a broader conception of the motivations behind intergenerational justice. Not everyone has descendants to worry about, but all, or almost all, individuals have interests concerning states of affairs that will or could come into existence after they are dead, and some of these interests are central to what makes lives and activities meaningful. These 'lifetime-transcending interests' are various. Many people have projects that they hope will make a contribution to people in the future; or they want their survivors to be able to carry on the work that they have begun. Others have ideals that they hope will be realized; some have communities, religious, political or ethnic, or families that they hope will flourish or people in the next generations that they care about. Some want their heirs to be able to enjoy their possessions, and their survivors to honour their wills and contracts. Others want a tradition to be continued, or a way of valuing to survive. And many are concerned about their posthumous reputation. As liberals they cannot *demand* that their successors take over their projects, value their contributions, realize their ideals, maintain their traditions and communities, or value an inheritance. But they will want future members of their society to be in the position to do so.

So in the manner of Rawls we can think of the requirements of inter-generational justice as being the result of a contract between the generations. In *Political Liberalism*, the generations are represented in the original position by a single generation which does not know where it falls within the sequence of generations, or whether it is rich or poor, any details about its predecessors or successors, or about its time in history.[13] Rawls says nothing about lifetime-transcending interests. But we can assume that members of this generation know that they are likely to have lifetime-transcending interests that are important to their lives. They will want to ensure that their successors are in the position to fulfil these interests: that they are able to choose to pursue the projects of their predecessors, maintain communities or carry on traditions. They will therefore want to perpetuate institutions that enable people to be self-determining and to pursue the objectives of their choice. They will want their successors to be capable of self-determination. They will therefore accept a principle of intergenerational justice that requires the maintenance of just institutions and social conditions which enable people to fulfil the requirements of these institutions.

This social contract story captures, in part, the reasons why we are able to describe as unjust a future world in which people have been rendered incapable of maintaining institutions that we would regard as just. It is unjust because people in it are not capable of appreciating and maintaining the kinds of things that we believe that the people of all generations ought to be in a position to value. To be sure, members of future generations may decide not to value these things. That is their right. But they should be able to live under institutions which give them the ability to appreciate their inheritance and the freedom to choose to accept it.

However, the story is inadequate in several respects. For one thing, it does not eliminate the possibility that our successors, who no longer value what we valued, take steps to ensure that their successors will not be able to value these things. Perhaps these successors have chosen to establish a theocracy and want to make sure that *their* successors are not capable of deviating from the true faith. The question once again arises of how can they be accused of committing an injustice when by their standards they are doing nothing unjust. Where do we get the authority to say that they would be wrong to do such a thing? Moreover, the social contract story, with its emphasis on the interests of people who are assumed to be present, is not able to take into account historical reasons that might exist for maintaining a tradition, or at least the conditions under which the tradition can be maintained.

Let us consider more closely some of the lifetime-transcending interests that people are likely to have. If I have been working on a project all of my life and want my designated successors to continue it, I cannot demand that they do so. But perhaps, at least in some circumstances, I can reasonably demand that they make an effort to appreciate my contribution: to understand what I was trying to do and why I thought it was important to make this contribution. Suppose that Mrs White, a businesswoman, put a lot of time, effort and some of the proceeds from her business into a programme for young people that was supported by community leaders and many of its members. She provided facilities for the use of the young and established a home that cared for needy children. Her aim was to provide a lasting contribution to the wellbeing of future generations. But shortly after her death, the community voted in new leaders who abandoned the programme for young people. The services that Mrs White built up were discontinued, the facilities sold off, and now no-one remembers Mrs White or what she contributed.

It is reasonable to judge that a wrong was done to Mrs White by her successors. Not by failing to continue her project. The community leaders did not have this duty. But it seems reasonable that she could have demanded of her successors, or others could have demanded on her behalf, that they make an effort to appreciate her contribution: what she did and her reasons for doing it. They should have understood and addressed these reasons. They should have been prepared to explain why their projects were superior.

Establishing just institutions in a political society is an intergenerational project. Individuals of past generations have often made great sacrifices to establish them, reform them, defend them and to make up for injustices of their society. They regarded themselves as labouring not just for their immediate successors but for generations further in the future. And since in the non-ideal world that we inhabit institutions are never perfectly just, the work of reforming and reshaping them must continue and will be handed on to succeeding generations. These future generations do not have a duty to continue the project. They may have other ideas about justice, or different ideas about institutions. But their predecessors could have reasonably demanded that they make an effort to appreciate what they accomplished: to understand why they thought their contribution was valuable.

If this is right then the people of each generation have duties in respect to the past as well as in respect to the future. Each generation ought to make an effort to appreciate the inheritance that they have acquired from their predecessors. Not just their immediate predecessors. The labour and sacrifices of past generations to establish just institutions can go back for many generations. But the existence of this duty means that each generation also has a duty to ensure that future members of their society will be in the position to appreciate the projects of their predecessors and to carry them on, if they choose. We cannot make a moral demand and then undermine the ability of future people to fulfil it. If a generation were to make it impossible for their successors to appreciate the intergenerational projects of earlier successors – whether through indoctrination, destroying information, or by genetic techniques – they would be doing an injustice. It would be an injustice to predecessors who were entitled to make lifetime-transcending demands, and it would be an injustice to those who will never be in the position to appreciate what their earlier predecessors laboured to provide for them.

Genetic enhancement and social progress

We ought not to allow genetic techniques to be used in a way that makes it difficult or burdensome for our successors to maintain institutions of justice or makes it impossible for them to appreciate the things that we value. But do we have a duty to lessen the burden: to bring about changes that will predispose people in the next generations to be more just? Hardy Jones implies the existence of such a duty by insisting that we have 'strong obligations' to create progeny who will have their rights respected and will respect the rights of others.[14] He thinks that we should avoid the conception of persons with a genetic endowment that makes them likely targets of rights violation, and also, presumably, we should not bring into existence those whose genetic endowment gives them a predisposition to violate the rights of others: for example, those who are prone to aggression or psychopathy. If this duty could be fulfilled then we would make it easier and less costly for future generations to maintain institutions of justice.

The conception of justice that I, and most other philosophers, defend does not give present people a duty of justice to make the burden to future generations of maintaining just institutions less than our own. If we have the duties that Jones endorses they must derive from a humanitarian concern to decrease the violations committed against individuals in succeeding generations, or to reduce burdens to future generations, or to make things better for them. Some of these conceptions of duty are problematic. For example, having black skin in some societies makes individuals a likely target of rights violations. But it would be wrong to deal with these violations by genetic techniques that ensure that no-one is born with a black skin. This would confirm the racist idea that having black skin is a defect, and it would probably not eliminate racism. It might decrease the ability of future people to maintain just institutions by decreasing their ability and inclination to tolerate differences. Moreover, it would make it difficult for these future citizens to understand and appreciate the ways in which Black citizens in past generations made a contribution to the reform of institutions of justice by fighting for their rights.

Jones's idea of duty is more acceptable if it means avoiding the creation of children with disabilities that predispose them to violate the rights of others – though there are familiar problems concerning how 'disability' is defined and how the objective is achieved. Perhaps we can do even better. Suppose that we develop genetic engineering techniques that enable us to increase the ability of members of the next generations to be good liberal citizens: to bring it about that they will be more inclined than most present people to be reasonable, far-sighted, tolerant and cooperative. Through this genetic engineering project we can make it much easier for our successors to maintain institutions of justice. Indeed, we can assume that they will be better able to maintain just institutions through difficult times. Liberals have often dreamed of bringing about a better, more just world through education. Why wouldn't it be just as acceptable to bring it about by genetic techniques?

It would, of course, be wrong to pursue such a project by illiberal means: by forcing people to have children of the 'right' kind, or by preventing them from having children who lack genetic improvements. We should also exercise a healthy liberal suspicion of governments who are able to wield, and potentially abuse, such great power. But there are further reasons for thinking that the project is misconceived. A world full of people like John Stuart Mill might be a good thing in some ways, but limiting in others. By producing characteristics that are supposedly good for a liberal society or by eliminating characteristics that are supposedly not so good, we would perhaps eliminate capacities that are good for other reasons.

Would the products of liberal eugenics be loving and caring as well as reasonable, tolerant and socially cooperative? Feminists have often criticized liberals for valuing characteristics that have traditionally been associated with men in the public realm. The imagined programme of liberal eugenics may turn out to be a way of making characteristics associated with masculinity more dominant in the population – perhaps to the detriment of activities essential for social life. Would products of liberal eugenics be prepared to sacrifice themselves for the sake of the survival of things that they value? A propensity to go to war for one's ideals is not generally a good thing, but it would also not be good if citizens were never prepared to make great sacrifices for their values and for the wellbeing of others. Would products of liberal eugenics be capable of a high level of artistic creativity? It could be the case that people with many different characteristics, including those with 'illiberal' propensities, are needed to maintain a just society and other things of value. This consideration is not a reason for rejecting altogether a programme of liberal eugenics, but it indicates that some attractive proposals for achieving a social good through genetic engineering could turn out to be dangerous and counter productive.

NOTES AND REFERENCES

1. J. Rawls, *A Theory of Justice* (Cambridge, MA: Harvard University Press, 1972) 287.
2. B. A. Ackerman, 'Justice over time' in *Social Justice in the Liberal State* (New Haven, London: Yale University Press, 1980) Part 2.
3. R. E. Goodin, 'Justice as reciprocity' in R. Elliot (ed.), *Environmental Ethics* (Oxford: Oxford University Press, 1983) 13.
4. R. M. Green, 'Intergenerational distributive justice and environmental responsibility' in E. Partridge (ed.), *Responsibilities to Future Generations* (Buffalo: Prometheus Books, 1981) 95.
5. M. L. Wissenburg, *Green Liberalism: The Free and Green Society* (London; Bristol; PA: UCL Press, 1998) 123, 126.
6. J. Habermas, *The Future of Human Nature* (Cambridge, London: Polity Press, 2003).
7. *Ibid.*, 82–4.
8. See N. Agar's presentation of these criticisms in *Liberal Eugenics: In Defence of Human Enhancement* (Oxford: Blackwell Publishing, 2004) 116–20.
9. D. Hume, *Enquiries Concerning Human Understanding and Concerning the Principle of Morals* edited by L. A. Selby-Bigge and P. H. Nidditch, 3rd edn (Oxford: Clarendon Press, 1975) 190–1.
10. H. G. Wells, *The Time Machine* (New York: Airmont, 1964).
11. A. Huxley, *Brave New World* (Harmondsworth: Penguin, 1955).
12. See for example, J. English, 'Justice between generations' (1977) 31 *Philosophical Studies* 91–104.
13. J. Rawls, *Political Liberalism* (New York: Columbia University Press, 1993).
14. H. Jones, 'Genetic endowment and obligations to future generations' (1976–8) 4 *Social Theory and Practice* 37.

8

Genetic preselection and the moral equality of individuals

David Neil

Suppose that it becomes possible to control the genetic traits of our descendants, and thus treat them as a product which can be engineered to our liking. Employing a Kantian vocabulary, Habermas says that this is a kind of intervention which should only be exercised over things, never over persons. In *The Future of Human Nature*,[1] Habermas develops a version of a common objection to genetic engineering – that it would involve treating humans as means rather than as ends. His formulation of this argument is important because he makes the novel claim that there is a somatic basis to our ethical freedom. We are embodied individuals and in order to regard ourselves as free and equal members of a community of similarly embodied individuals, we need to stand in a certain relationship to our own unchosen physical characteristics. The prospect of choosing the positive genetic characteristics of another person threatens to change the nature of that person's relation-to-self in a way that undermines his or her potential to become fully autonomous.

A number of philosophers working within the liberal tradition have argued that, for certain purposes, genetic selection and enhancement of embryos may be consistent with liberal principles. Liberal eugenics distances itself from the dark history of authoritarian, state-directed eugenics programmes, but asserts that parents' rightful freedoms entitle them to pursue some eugenic goals with respect to their children.

For instance, John Harris points out that we accept the conditioning of children by education, towards the development of intelligence, fitness and so on. Where the goals of education and genetic enhancement are the same, it is prima facie inconsistent to embrace one mode of causal determination while rejecting another.[2] John Robertson argues that a right to use genetic screening and enhancement technologies can be derived from the rights of procreative liberty that parents already enjoy.[3] Buchanan, Brock, Daniels and Wikler have

The Sorting Society: The Ethics of Genetic Screening and Therapy, ed. Loane Skene and Janna Thompson. Published by Cambridge University Press. © Cambridge University Press 2008.

argued that genetic interventions to prevent disabilities are obligatory on grounds of justice and duties to prevent harm; and that genetic enhancements are permissible when they are pursued within certain constraints of liberal justice.[4] Nicholas Agar has argued for a 'liberal eugenics' which allows parents to select genetic 'goods' for their children, provided engineered enhancements respect a 'eugenic difference principle' of Rawlsian inspiration.[5] In a similar vein, Fritz Allhoff has argued that germ-line genetic enhancements are permissible where they serve to augment Rawlsian primary goods.[6] For all of these authors, the permissibility of genetically sorting and programming embryos is qualified in various ways. Nevertheless there is a good deal of support for the view that, in liberal societies, some kinds of embryo engineering should be allowed in the name of freedom, and perhaps even promoted in the name of equality.

Opponents of genetic engineering often appeal to religious doctrines, or derive an injunction against genetic tampering from a 'thick' and morally loaded account of human nature. Conservatives of various colours offer a range of moralized conceptions of the 'human' and its proper limits. For such critics, technologies that promise a 'post-human', 'transhuman' or 'super-human' future are all just roads to the inhuman, and thereby to perdition. A central principle of liberalism, however, is state neutrality with respect to the good. In a pluralistic liberal society, policies and prohibitions binding for all citizens cannot be premised on substantive moral views that only some of the citizens accept.

It is for this reason that Habermas's argument against genetic engineering is particularly interesting. His argument also depends on a certain moral image of human nature or, as he puts it, an 'ethical self-understanding of the species'.[7] However Habermas constructs his case within the normative framework and resources of *political* liberalism. Political liberalism accepts moral pluralism as a social fact and seeks to define the minimal moral consensus necessary for peaceful coexistence in a pluralistic society. Central to the idea of a liberal society is a commitment to the formal moral equality of citizens, standardly conceived in terms of a universalized legal and moral respect for individual autonomy. Habermas wants to show that allowing parents to design the genetic traits of their offspring, according to their subjective preferences, threatens to undermine the social preconditions of liberal equality.

A society of equals is one whose members possess equal rights and are bound by reciprocal duties. The members of the community place one another under moral obligations and expect each other to conform to norms of behaviour. For a moral agent, norms are not mere rule following. A well-trained dog may be reliably obedient and entirely predictable in response to the commands of its master, but it is not a moral agent. For rational moral agents, the normativity of shared social rules derives from their understanding and acceptance of those rules. They submit to norms voluntarily because they comprehend those norms as derived from shared reasons that the members of their community have to regulate their behaviour with regard to each other. Equality implies an essential symmetry in the relationship of moral agents, which requires that normative claims take the form of *universalized* reasons. The notion of 'human dignity' as inviolable is closely connected with this symmetry. To violate another person is to violate one of his or her basic rights, and the moral force of rights is ultimately derived from the mutual recognition that we possess shared reasons.

At this point Habermas introduces the body into the explanation of morality. It is in virtue of our rational nature that we are capable of being moral agents, but it is in virtue of our physical embodiment that we *need* to become moral agents. A community of disembodied intellects would have no use for morality. Habermas affirms a constructivist view of the origins and function of morality. Morality is a socially constructed solution to the very material problems presented by our mutual dependence and vulnerability. He puts it thus:

I conceive of moral behavior as a constructive response to the dependencies rooted in the incompleteness of our organic make-up and in the persistent frailty (most felt in the phases of childhood, illness and old age) of our bodily existence. Normative regulation of interpersonal relations may be seen as a porous shell protecting a vulnerable body, and the person incorporated in this body, from the contingencies they are exposed to.[8]

We need morality because we are dependent on each other and we can harm each other. Autonomy is not an abstract property of rational beings. To have autonomy is an achievement; a fragile, easily destroyed achievement, which is only preserved as long as individuals are mindful of their physical vulnerability and mutual dependence. We realize autonomy, as a meaningful freedom to be the authors of our lives, only through being equal members of a community of moral agents.

For Habermas the world of morality is a symbolic network constituted by the relations of mutual recognition of communicatively acting persons. Moral recognition is about the giving and receiving of reasons. The development of personhood substantially is the process of identifying oneself both as a person in general – that is, as a member of a community of persons – and simultaneously as a unique individual who is morally non-exchangeable. In order to assist the socialization process we anticipate it in the neonate, and extend to infants many of the rights and protections due to persons. Habermas contends that this attitude to infants – which inscribes them socially and legally as persons before they are in fact persons – shows that human life *as such* has moral value.

Habermas notes that the fetus and the neonate are already the subject of an 'anticipatory socialization'. Our moral responsibilities towards children are structured by an anticipation of their future autonomy. A standard test of whether parental authority has been exercised well or badly is to consider how decisions would be justified to the later adult. The baby will one day be a grown person, who will ask her parents: 'why did you make me do X, or forbid me from Y' and they must give an explanation. Parental authority is like the authority of a trustee, who does not own what he holds in trust. The developing child already belongs to the moral world of reason giving in the sense that parents should behave *as if* they are answerable to the future adult, who will have a right to a reasonable justification of the way that parental power was exercised.

For Habermas, the permissibility of prenatal, therapeutic gene manipulation is determined by the reasonableness of assuming 'virtual' consent, when actual consent is not possible. Liberal eugenicists have challenged the distinction between genetic therapy and genetic enhancement, on the grounds that conceptions of disease are dependent on social norms and values. Habermas insists that we do need to draw a line between therapy and enhancement, as marking a boundary between acceptable and unacceptable genetic interventions. Genetic manipulations may only be classed as medical cures, rather than biotechnological enhancements, when we have broad social consensus that the condition to be 'corrected' is a disease. An 'assumed consensus can only be invoked for the goal of avoiding evils which are unquestionably extreme and likely to be rejected by all.'[9]

The very possibility of a shared consensus, achieved through public reason, presupposes that citizens recognize each other as reasonable beings. The condition that Habermas calls 'post-metaphysical thought' is the condition of culturally, religiously and socially diverse societies. In these liberal pluralistic societies the just is prior to the good. In both theistic and naturalistic ethical theories, the good is typically grounded in a metaphysical picture of the essential nature of humans and values. Because no single moral metaphysics commands the field, *public* reason is 'post-metaphysical' for reasons of pragmatic necessity. Since any account of the good life is unacceptable to those who subscribe to an alternative conception of the good, diverse societies are faced with the alternatives of mutual tolerance or intractable conflict. Thus in Rawls's account, for instance, the demands of justice are given by the principles of fairness that protect the separateness of persons. However Habermas insists that:

> ... this 'priority of the just over the good' must not blind us to the fact that the abstract morality of reason proper to subjects of human rights is itself sustained by a prior *ethical self-understanding of the species*, which is shared by all *moral persons*.[10]

Liberal tolerance itself does not make sense unless we conceive of human beings as animals capable of developing an autonomous morality. From the liberal perspective, the various beliefs about the good that we observe in pluralistic societies deserve respect only in so far as we imagine that these alternative ways of life can be autonomously chosen. Liberal neutrality, then, does not mean that liberalism presupposes no ethical image of the human. Liberalism must picture humans as possessed of whatever characteristics are necessary for understanding ourselves as 'ethically free and morally equal beings guided by norms and reasons'.[11] In traditional moral debates between established schools of thought these assumptions can sit far enough in the background as to appear self-evident. Today, as previously unimagined possibilities for the manufacture of humans appear on the horizon of feasibility, the assumptions underwriting this 'ethical self-understanding of the species' are suddenly brought to the surface. Individual characteristics which have always been outside the scope of parental choice may become design options in a new order of technologically mediated procreation. What if some of these interventions impact negatively on a 'designed' person's ability to see him- or herself as an autonomous and morally equal being? A consistent liberalism cannot countenance genetic interventions that might impair the development of autonomy in the later person.

Habermas claims that one of the preconditions for being able to conceive of ourselves as both the authors of our own lives, and as equal members of the moral community, is that we enter the world as beings that are grown and not made. He develops this claim through a critique of liberal eugenics. Habermas quotes a passage from Nicholas Agar's essay 'Liberal Eugenics', in which Agar states that 'the distinguishing mark of the new liberal eugenics is state neutrality'.[12] Parents will select 'improvements' in their children according to their own values. Where authoritarian eugenics would restrict ordinary procreative freedoms liberal eugenics proposes radical extension of those freedoms.

Liberals can only admit eugenic modifications if the genetic modifications do not limit the autonomy of the future person or restrict his or her ability to interact with others as an equal. Thus liberal eugenicists emphasize the similarities between genetic enhancement and other interventions already allowed to parents. As Robertson puts it:

If special tutors and camps, training programs, even the administration of growth hormones to add a few inches in height are within parental rearing discretion, why should genetic interventions to enhance normal offspring traits be any less legitimate?[13]

Advocates of liberal eugenics emphasize the complex interactions of genes and environment in determining an individual's traits. If the causal influence of genes and environment are of the same order, we should regard genetic and environmental interventions as morally on a par also.

Liberal eugenicists who insist on the parity of genetic enhancement and education fail to acknowledge that this comparison 'cuts both ways', so to speak. Children who go to well-resourced, privately funded schools, instead of poorer public schools, typically translate that luck into opportunity and wealth. In most Western societies parents de facto enjoy the right to allocate any proportion of their resources to their children's betterment. The exercise of that right produces social inequalities that are manifestly unjust. If eugenic technologies are allocated by the market, in a world not too economically distant from ours, then we should expect that such technologies will be used by the wealthy to further guarantee the class privileges of their children. From the perspective of distributive justice, the similar effectiveness of genetic and educational enhancements might count *against* parental freedom to pursue eugenic goals. However our concern here is not with the legitimacy of gene technology markets, but with the impact of genetic preselection on *autonomy*.

For Habermas the comparison of a prospective eugenic freedom with parental discretion in respect of education is a 'dubious parallel' which 'presupposes a leveling out of the difference between the grown and the made'.[14] His objection rests on a phenomenological distinction between 'being' and 'having' a body. In *The Future of Human Nature* this distinction is inadequately described, but for the purposes of this discussion we will take the essential point to be this: in order to be a moral agent who takes responsibility for my life and actions, I need a certain kind of subjective relationship or attitude towards my own body. It is my body that experiences, feels and acts. It is with reference to the body that we distinguish self from other; active from passive; what is done by me from what is done to me. In order to experience my thoughts and actions as my own, I need to identify with my body in the right way. To be autonomous I must be 'at home', so to speak, in my own skin. There are a number of conditions in which one's own body or mental states are experienced as alien and subject to external determination, and in such a state one's autonomy is severely compromised.

A crucial condition of autonomy is that I have a measure of agency over myself. I can take an objective and critical stance towards my own dispositions and behaviours, my personal history and to some extent my body. I can take myself as a project and attempt to change what I am. Causal explanations of phenotype refer to complex interactions between genes and environmental factors. Liberal eugenicists stress this point to argue for the moral equivalence of shaping persons by genetic or environmental design. Habermas, however, thinks that a person who has been genetically programmed might not be able to integrate the knowledge of that programming into a successful adult perspective on his or her life. Such a person cannot effectively accept or reject the values expressed in the selection of his genes, in the way that he might come to reject the values espoused in his education.

Some of Habermas's liberal critics have misunderstood the point that he is making here. They interpret Habermas as appealing to the apparent impossibility of rebelling against one's own genes in order to establish a clear moral difference between genetic and environmental enhancements. They charge that Habermas's attempt to draw a sharp distinction between genetic and environmental interventions – the former being unrevisable and the latter revisable – reveals his simplistic and erroneous genetic determinism. David Wasserman presents this criticism as follows:

Why should Habermas believe that it is any more hopeless to be at odds with the 'genetically fixed' than the 'environmentally fixed' intentions of a third person? To the extent that parents shape the character and abilities of their already-born children, they do so largely at a time when those children are too young to contest their influence in any coherent or effectual way.[15]

Replying to Habermas's criticisms, Nicholas Agar quotes this passage with approval and, also in agreement with Wasserman, suggests that Habermas's objection hinges 'not on the irresistibility of genetic interventions, but instead on their unilateral nature'. Agar suggests that we might view genetic enhancement as less unilateral if we bear in mind how genes influence traits.

It is true that someone cannot, as an adolescent, change the fact that parents have engineered his genome – but he does have some control over the *influences* of the genes that he has received. Many genetic influences on complex traits such as intelligence take place after a person's birth, and they require a specific environment to have the effects that geneticists associate with them. The adolescent has a say in whether the influence of the inserted gene gets matched with the environment without which it cannot have its effect.[16]

This response misses the mark. The term 'unilateral' describes decisions or actions imposed on one party by another, without consent. 'Unilateral' does not mean that the subject of a unilateral action has no means of resistance. An attempt to control another person is not more unilateral if it succeeds and less unilateral if it fails. The wrong of acting unilaterally is the wrong of arrogating to oneself a power of decision that disregards the legitimate claims of others. Even in the ordinary course of child-rearing, parents can coercively shape their children in ways that are damaging for the development of autonomy. Habermas's objections to genetic enhancement cannot be refuted by pointing out that some environmental interventions may be equally irreversible from the perspective of the designed person. Nor can one reply by pointing out that genetic enhancements may fail to realize their intended effects in an uncooperative subject. Habermas draws our attention to the limits that respect for autonomy imposes on the *kind* of intentions we may have with regard to others. Genetic enhancement enables parental values and intentions to enter into a domain of the child's life history from which they have hitherto been absent, and where there can be no reasonable assumption of retrospective consent.

If we see ourselves as moral persons, we intuitively assume that since we are inexchangeable, we act and judge *in propria persona* – that it is our own voice speaking and no other. It is for this 'capacity of being oneself' that the 'intention of another person' intruding upon our life history through the genetic program might primarily turn out to be disruptive.[17]

Habermas claims that in order to successfully regard ourselves as the authors of our own actions and aspirations, we need an origin that, as he puts it, 'eludes human disposal'. For each of us our birth is a natural point of origin that precedes our socialization. Babies are nature without culture, and thus the newborn is the definitive symbol of the possibility of new beginnings. My birth may have been the consequence of my parents' decision to have a child, but the particular body that I am was mostly outside their control – a permutation in nature's lottery.

For Habermas this undesigned, pre-social origin that we all share is a precondition of our moral equality. The domain of morality is the domain in which we may demand of others justification for actions that affect us. I respect the autonomy of another person when my actions towards that person are regulated by the general norms that he or she, as a rational agent, also accepts and is bound by. When conflict arises, respect for the humanity of others requires that we engage in reasonable discourse to find norms that can elicit consensus. As noted above, our treatment of neonates must satisfy a reasonable presumption of retrospective consent. When we are dealing with a baby who is not yet a person, this hypothetical consent can only be assumed if life-shaping decisions are guided by generally accepted norms. If I design a child to satisfy my own idiosyncratic preferences, then I cannot assume the retrospective consent of the later adult. The relationship between designer and genetically designed is permanently asymmetrical.

Our bodies have a vast number of characteristics with regard to which we are variously happy, unhappy or indifferent. For instance, I have a form of myopia that is, probably, genetically determined. It is correctable with glasses, but it can be annoying and I would prefer not to be myopic. The explanation of my myopia does not refer to any decisions made by other people. Nobody is responsible for the presence, in my genome, of genes that caused my eyeballs to grow too long. Because my myopia is not the result of the actions of another person it can raise no *moral* complaint.

By contrast I can make moral judgements about my education. As an adult I can take a critical and revisionist stance towards the values and beliefs that were espoused in my education, and I can attempt to undo some of the effects of that education on my dispositions and behaviour. With regard to the different aspects of my socialization the practical difficulties of refusal or revision vary greatly. From the perspective of autonomy, what matters is that the effects of my education are not inaccessible to me – I can get some purchase on these features of myself.

Imagine that a brain programming technology exists that allows educators to create permanent attitudes and beliefs. This technology can install indelible first-, second- and even third-order desires into a person's psychology, permanently eliminate countervailing desires and install all of the beliefs necessary to make these desires seem rationally compelling. The subject of this treatment is incapable of ever revising these desires, *wanting* to revise them, or even entertaining the idea that he or she might have reason to revise them. Imagine a proposal for the beneficent use of this technology to inculcate in children desires to eat only nutritious food and lifelong disgust towards junk food and cigarettes. Whatever behavioural benefits might be achieved with such a technology, for liberals, its use would be an intolerable violation of autonomy. It would be no argument for such methods to point to the fact that we currently educate children about health and diet, and brain programming would more effectively achieve the very same objectives. Note that the problem here is not just that the subject of this treatment cannot revise the programmed attitudes. The problem is that desires installed in such a fashion can never be autonomously held. The programmed can never evaluate the normative decisions that determined their psychology and so their programmers have made them into something less than moral equals. In shaping future adults, the obligation to promote autonomy in children and adolescents impose limits not just on the goals, but also on the methods by which developmental goals may be realized.

A decision to select the genetic traits of another person is, like most decisions affecting others, an object of moral evaluation. To express one's personal values in the genome of one's child is unacceptable under the principle of respect for autonomy. The objection from autonomy does not rest on empirical assumptions about genetic determinism; nor on assumptions about the respective contributions of genes and environment in determining phenotype. The problem is with the *intent* of such a project. If I genetically design a future person in accordance with my desires, where subsequent consent cannot be assumed, then I have treated that person as a means and not as an end in him- or herself. I have manufactured the body of another person as if it were something I owned.

An argument commonly raised in support of pre-implantation genetic diagnosis and prenatal genetic enhancement is that no-one is harmed. The genetic intervention is intrinsic to the identity of the later person, such that the designed person cannot object to the designer's actions without implying that he or she should not exist at all. Is there any coherent sense in which one can feel violated by learning of a parental decision that was a precondition of one's existence? Nicholas Agar argues that 'there is some reason to think that enhancement technologies will forever elude Kantian morality's focus'.[18]

The principle of respect for autonomy tells us nothing about the claims of merely possible individuals, with respect to the actions which bring them into existence. Can Kantians, for instance, object to the intentional creation of non-rational humans, who could then be used instrumentally? The manipulation of pre-personal human life seems to escape deontic constraints on the treatment of persons.

The right liberal response here is to point out that objections to eugenic enhancement do not have to be presented from the perspective of the subjects of such enhancement. It is not a matter of speaking for a silent 'victim'. Habermas's concern is about the conditions for the survival of liberal societies. Liberal societies must, by definition, be communities of moral equals. Inscribing our preferences into the genome of our descendants might undermine the preconditions for the moral equality of persons. Considerations of equality constrain how we may project our values into the future. If our children are to become our equals then they must be able to autonomously affirm, revise or reject the values we bequeath. In fact, the danger for liberalism arises precisely *because* the Kantian principle of respect for rationality has no application to choices between merely potential persons. Genetic programming decisions affect others and thus demand moral justification; yet these decisions are taken in a domain where rules of reciprocity between equals can find no traction.

Is prospective life opportunity a legitimate criterion for deciding the type of people who should exist? The liberal debate over genetic enhancement reveals a surprising tension between the formal constraints of moral equality and the goals of liberal justice. Liberal theories of distributive justice are standardly concerned with mitigating the undeserved effects of luck on opportunity. Yet Habermas is claiming that natural birth is a *precondition* of autonomy. Nature's lottery is certainly not just. The very paradigm of undeserved disadvantages, which liberal justice would ameliorate, are unhappy accidents of birth. For advocates of liberal eugenics, genetic diagnosis and engineering appears to offer the possibility of more effectively addressing important sources of injustice. Rather than merely compensating those who lose the birth lottery, we might better reduce unfairness by making birth less of a lottery. Agar, for instance, argues that genetic interventions are acceptable when they enhance the child's capacity to realize his or her *self-chosen* life plans, and when the allocation of genetic 'goods' satisfies Rawls's difference principle.[19]

It is true that one's own body can be experienced as inadequate and limiting. To enjoy meaningful opportunity our physical and mental powers must be adequate to our plans. In that sense an individual's life opportunities can be genetically enhanced. But autonomy is not merely a function of one's performance capacities. As Habermas reminds us, autonomy is a relational property. To be autonomous I must be able to see my intentions and actions *as mine*, in contradistinction to the intentions of others. Our moral sense of independence and responsibility for our actions depends on believing that we are the *authors* of our plans. If we write our values into the genome of our children, it is far from clear that the boundaries of authorship, and thus autonomy, can be preserved.

NOTES AND REFERENCES

1. J. Habermas, *The Future of Human Nature* (Cambridge: Polity Press, 2003).
2. J. Harris, *Wonderwoman and Superman* (Oxford: Oxford University Press, 1992) 140–2.
3. J. Robertson, *Children of Choice: Freedom and the New Reproductive Technologies* (Princeton, New Jersey: Princeton University Press, 1994) 166–7.
4. A. Buchanan, D.W. Brock, N. Daniels and D. Wikler, *From Chance to Choice: Genetics and Justice* (Cambridge: Cambridge University Press, 2000) 302.
5. N. Agar, 'Liberal eugenics' (1998) 12(2) *Public Affairs Quarterly* 137–55.
6. F. Allhoff, 'Germ-line genetic enhancement and Rawlsian primary goods' (2005) 15(1) *Kennedy Institute of Ethics Journal* 39–56.
7. J. Habermas, *The Future of Human Nature*, note 1 above, 40.
8. *Ibid.*, 33.
9. *Ibid.*, 43.
10. *Ibid.*, 40.
11. *Ibid.*, 41.
12. N. Agar, 'Liberal eugenics', cited in J. Habermas, *The Future of Human Nature*, note 1 above, 49.
13. J. Robertson, *Children of Choice: Freedom and the New Reproductive Technologies*, note 3 above, 167, cited in J. Habermas, *The Future of Human Nature*, note 1 above, 49.
14. J. Habermas, *The Future of Human Nature*, note 1 above, 50.
15. Cited in N. Agar, *Liberal Eugenics: In Defence of Human Enhancement* (Oxford: Blackwell Publishing, 2004) 117.
16. *Ibid.*
17. J. Habermas, *The Future of Human Nature*, note 1 above, 57.
18. N. Agar, *Liberal Eugenics: In Defence of Human Enhancement*, note 15 above, 43.
19. *Ibid.*, 136–7.

9

Genes, identity and the 'expressivist critique'

Rob Sparrow

Introduction

Technologies such as prenatal testing, combined with the option of abortion, and pre-implantation genetic diagnosis now give prospective parents unprecedented power to choose the genetics of their children. In effect, they allow parents to sort embryos according to whether they have desirable or undesirable genes. A society in which such technologies become widespread – as they have in many industrialized nations already – might be thought of as a 'sorting society'.[1] This description, however, immediately draws attention to another, more disturbing, potential in these technologies. Critics of the sorting society worry that it involves choosing between different 'sorts' of people, deciding who will be born on the basis of a belief that some sorts of people are better than others.

The shameful historical legacy of racial eugenics has meant that there is little open enthusiasm for using modern technologies of genetic selection to select for (supposed) racial traits. While technologies enabling sex selection have been widely adopted to that purpose, their use in this fashion has, I think, at least as many critics as admirers amongst those writing about the ethics of this practice. Instead, these sorting technologies have been taken up and defended most enthusiastically in the service of the goal of preventing the birth of children who might suffer from various disabilities. As a result, it has been critics from within the disability community who have thought hardest about – and have raised some of the most forceful objections to – the development of the sorting society. Critics including Adrienne Asch,[2] Susan Wendell,[3] Marsha Saxton[4] and Deborah Kaplan[5] have argued that in using technologies of genetic selection to ensure that children are not born with disabilities we may express a disrespectful attitude towards existing persons with disabilities.

The Sorting Society: The Ethics of Genetic Screening and Therapy, ed. Loane Skene and Janna Thompson. Published by Cambridge University Press. © Cambridge University Press 2008.

In this paper, I explore this 'expressivist critique'.[6] I begin by setting out the expressivist critique and then highlighting, through an investigation of an influential objection to this critique, the ways in which both critics and proponents of the use of technologies of genetic selection negotiate a difficult set of dilemmas surrounding the relationship between genes and identity. I suggest that we may be able to advance the debate about these technologies by becoming more aware of the ways in which this debate is itself in part a political contestation over this relationship.[7] Ultimately, I will argue, the real force of the expressivist objection lies in its capacity to draw our attention to *political* questions about the role of the state and about relationships between different social groups rather than between parents and prospective children. That is to say, crucial issues, when evaluating the force of this criticism, turn out to be: the nature of the institutions which determine how decisions about prenatal selection are made; and how we think of each other, that is, what we take to be the defining characteristics of human beings. Paradoxically, arguments about the ethics of the 'sorting society', both supportive and critical, are an important arena in which these institutions and these ideas about identity are contested and shaped. An increased awareness of the reflexive nature of the process of debating these issues may assist us in better negotiating them.

The expressivist critique

An important criticism of the project of the 'sorting society' derives from a concern that selecting against embryos with genetic disorders, or terminating pregnancies on the basis that they are likely to result in the birth of a child with a disability, expresses morally reprehensible negative attitudes towards people with disabilities. There are in fact two different messages that the decision to choose one embryo over another might be taken to express.[8] Firstly, the decision to terminate a pregnancy – or not to select an embryo for implantation – on the basis of a diagnosis of a genetic condition contributing to a disability might be thought to send a message along the lines of 'it would have been better if you had never been born'.[9] Secondly, the decision to choose an embryo which is more likely to become a child without a disability rather than one which might develop into a child with a disability might be thought to imply that those doing the selection believe that people without disabilities are superior to those with disabilities.[10] The use of technologies of genetic selection to prevent disability may therefore convey both a message of superiority over, and of lack of respect for, those living with disabilities.

This objection has been dismissed by many philosophers who have examined the topic on the basis of a distinction between (dis)respect for persons and our attitudes towards disability.[11] A desire that a child not be born with a disability which could be avoided through the use of a technology of genetic selection need not express any attitude towards existing persons with disabilities. Parents may simply believe that it is better to be born without a disability rather than with a disability, without having any feelings about the *worth* of the lives of those with disabilities. According to this way of thinking, choices made on the basis of beliefs about the negative impacts of disabilities need not express any message about the worth of the lives of those currently living with disabilities.

The question of what sort of actions and utterances express evaluative attitudes, what they express, and what determines this, is a complex and controversial one. I will return to this matter in a later section of the paper. Firstly, however, I want to consider an argument which draws on the fact that both genetics and environment play a role in shaping the ultimate character of an organism in order to suggest that we are obligated to make use of technologies of genetic selection to improve the opportunities available to those who will be born. This argumentative strategy is a familiar one in debates about 'liberal eugenics'[12] but I want to rehearse it here because I think presenting it in the stages set out below can usefully illuminate just what is at stake in arguments about the ethics of the sorting society and, in particular, the way these arguments raise difficult philosophical and – I will argue – political questions about personal identity.

Genotype and phenotype

A compelling chain of analogies appears to demonstrate that an obligation to avoid – or even terminate – particular pregnancies on the basis of a diagnosis of a genetic disorder follows naturally from an obligation to treat injuries sustained after birth which we all ordinarily acknowledge.[13]

Few would be prepared to argue that we should not use medical technology to heal and correct injuries or conditions which have occurred after birth. Yet the purpose of such interventions is to prevent a disability. In this case, however, it seems wildly implausible to suggest that we should not proceed with surgery because to do so would be to express a negative attitude towards people with disabilities. This argument would rule out any medical treatment of conditions which, if left untreated, might lead to disability. It seems clear in this case that, on the contrary, we are obligated to proceed with such surgery if it offers any chance of success.

If such postnatal surgery is acceptable, however, it is hard to see what would be wrong with surgery performed in utero to avoid a condition which might otherwise negatively affect the child. The only difference between such surgery and an operation performed postnatally to the same effect is a matter of a few weeks and the location of the child being operated upon. It is far from clear that these factors should make a difference in the ethics of the procedure.

Similarly, if surgery of this type is acceptable it is difficult to see what would be wrong with making use of a (hypothetical) somatic cell gene therapy to achieve the same effect.[14] In theory, an efficient somatic cell gene therapy might introduce new genes into the somatic cells of the developing embryo to correct for the effects of defective genes. In terms of its effects on the eventual capacities of the infant, this appears only to move the moment at which the condition is corrected a few months earlier in the course of the pregnancy.

However, at this point intuitions about any obligation to carry out such gene therapy begin to diverge. As I will discuss further below, rather than altering the capacities of the child that is eventually born, such gene therapy might be thought to result in the birth of a different child – with dramatic implications for the ethics of the procedure.

The intuition that genetic interventions may affect the identity of the child that will be born is, I think, even stronger in the case of a hypothetical *germ line* gene therapy. Such a therapy would introduce new genes, which would replace (or repair the effects of) the genes that would otherwise result in a disability, into the chromosomes of every cell in the developing embryo. The eventual presence of the new genes in the gametes of the developed infant means that these changes could be passed on to future generations. As a result of such therapy, the embryo will now have a different genome.

This method of intervening to prevent a child being born with a disability can be described in two different ways. It can be described as altering the character of the (same) organism so as to change the phenotype of the infant,[15] which renders it analogous to the in-utero surgery described above. However, because it involves altering the genome of the embryo, in so far as many people have the intuition that genetics play a role in individuating organisms and on the, not unreasonable, assumption that a different organism means a different person, germ line gene therapy might be argued to involve changing who will be born; that is, that this kind of intervention secures the birth of a different person.[16] This way of understanding this scenario suggests that it is more analogous to the case of pre-implantation genetic diagnosis (PGD).

If germ line gene therapy is acceptable then it is difficult to see what would be wrong with using PGD (or prenatal testing and selective termination of pregnancy) instead, in order to ensure that the child which is born will have the appropriate genotype. Indeed, we could imagine that the genetic selection and the germ line gene therapy might lead to the birth of a child with the very same genome. That is, if we imagine a gene therapy which alters an embryo with a genotype, 'A' (which includes a gene coding for a defective protein), so that it will instead have a revised genotype, 'B' (which will lead to the embryo developing normally), then administering this gene therapy prevents a child being born with genotype A and instead ensures a child with genotype B will be born. The very same outcome could be achieved by employing PGD instead to choose between two embryos, one of which has genotype A and one of which has genotype B. In each case we have an action that determines the genetics of the child which will be born. However, whereas the gene therapy might be thought only to alter the character of the embryo, PGD seems to involve choosing a different organism – and therefore, presumably, a different person – to be born.

Because the phenotypic outcome at the end of all these interventions is the same it is, at first sight at least, difficult to explain what grounds we could have for accepting some of them but not others. Indeed, given that most people accept that we are obligated to initiate surgery in the case of the accident immediately after birth, it is difficult to explain why we are not similarly obligated to use PGD.[17]

There are two obvious ways in which this conclusion may be resisted.

Firstly, it may be pointed out that when prenatal testing and selective abortion are employed to shape future persons this involves the destruction of the unwanted embryos, which is not the case with the other technologies. Yet many of those writers drawn to the expressivist objection deny that the termination of pregnancies (or the destruction of embryos) per se is morally problematic.[18] This suggests that it is not the destruction of embryos itself that is the issue here but rather their destruction on the basis of a particular character trait. Thus Adrienne Asch[19] has argued that it is the destruction of an embryo on the basis of a diagnosis of a genetic condition that might lead to a disability which communicates a negative attitude towards people with disabilities. When parents choose to terminate a pregnancy on the basis of a diagnosis of a possible genetic disorder in the fetus or embryo, the developing life that they had previously celebrated and welcomed into their lives is instead suddenly and emphatically rejected. The extent to which any such attitude towards a potential life must extend to, or communicate something about attitudes towards, actual persons will be discussed further below.

Secondly, while all these interventions result in the child that is born having the same phenotype, PGD and (perhaps) germ line gene therapy (and perhaps even somatic cell gene therapy) achieve this by bringing about the birth of a *different* child. One set of therapies improves the life of the child who is eventually born whereas the other set of interventions alters who comes to be born. Thus it might be argued that the use of these technologies involves asserting that one embryo is better than another and therefore, perhaps, that people of one sort are better than another.[20] Again, it is not implausible to suggest that there is something about a course of action that must necessarily alter the whole character of the future person, which is problematic in a way in which courses of action that merely alter one (or perhaps several) particular character traits are not. In particular, such action seems to involve a comparison between different persons (or, at least, embryos) that might need to be based on comparison between different sorts of lives.[21]

The dangers of genetic determinism

However, in linking such decisions to attitudes towards persons with disabilities both these ways of resisting this chain of analogies seem to involve conceding that the identity of the person who would be born – what sort of person they are – will primarily be determined by their genetics.[22] The expressivist critique therefore seems to involve a genetic determinism that has been widely criticized elsewhere in bioethics.

People are, as we must continually remind ourselves in debates about genetics, more than their genes. This is true as a matter of developmental biology. Genotype only produces phenotype in a given environment. Environmental conditions are thus equally important as genes in determining an organism's character.[23] However, more importantly, it is also true in relation to our duties to others. The respect we owe to other persons is owed to them regardless of their genetics. Whether or not a person has a genetic condition leading to disability is irrelevant to the moral respect that is owed to them. It seems odd, therefore, for critics writing from a disability perspective to choose genetics as what distinguishes one person from another.

Moreover, in suggesting that steps taken to reduce disability express disrespect for persons with disabilities, the expressivist critique seems to imply that persons with disabilities are essentially constituted by their *disabilities*.[24] Consider the case of a couple who terminate a pregnancy because, if they proceed with it, they would have four children, when they only intended to have three. In this case we are unlikely to conclude that they express a disrespectful attitude towards all those people who are the fourth child in the family.[25] In order to establish that selection against embryos expresses disrespect for an existing group of people, it seems that we must hold that there is a certain sort of person who is characterized by sharing the trait on the basis of which the embryos are being selected against. Being a 'fourth child' does not pick out a sort of person who might claim to be the object of disrespect or hostility expressed by parents who make such choices. For the expressivist critique to function, however, it must be the case that disability does.[26] This claim is obviously problematic in the context of a critique from a disability rights movement that is elsewhere insistent that persons with disabilities are persons who happen to have disabilities rather than disabled persons.[27]

The manner in which the expressivist critique relies on a perverse claim about the identity of people with disabilities is highlighted by Mary Ann Baily when she observes, of critics of the use of technologies of genetic selection to screen out disabilities, that,

Their picture is of a line of babies waiting to be born, and a quality control officer coming along and throwing "people like them" out of the line so they never make it to earth.[28]

That embryos might be 'people like them' requires both that people with disabilities are appropriately interpellated with reference to their disabilities and that the character trait of 'being a person with a disability' inheres in the genes.[29]

However, it would be a mistake to think that it is only critics from the disability community who are inclined to individuate and draw conclusions about embryos – and therefore the beginnings of persons – on the basis of their genes. The advent of modern genetics has led to social discourses about identity becoming dominated by 'gene talk'.[30] Since the discovery of DNA, race, sexuality, sex, and gender have been re-imagined as 'genetic conditions'.[31] This process has accelerated further since the initiation and subsequent completion of the Human Genome Project.[32] As a result, it is increasingly the case that a wide range of character traits are thought of as coded 'in the genes'.[33] The use of PGD only makes sense if we can say something about the sort of life a person is likely to live purely on the basis of their genetics.

Moreover, I noted above that many people are inclined to think that PGD and perhaps germ line gene therapy result in a different person being born rather than just altering the character of the (same) person who is born. The temptation to link genetic identity and continuity to personal identity across the early stages of life is a product of our scientific understanding of the role played by genes in the development of an organism and our sense that the human organism begins at conception. Yet this way of individuating human lives has perverse and puzzling implications for the description and ethics of the use of technologies of genetic selection. It has the consequence, noted above, that technologies of genetic selection change who comes into existence rather than alter the character of any particular person.[34] As the persons who are brought into existence using these technologies would not have come into existence except for the use of these technologies they are unable to claim that they would have been either better or worse off if these technologies had been used otherwise.[35] As a result, it seems that the use of these technologies does not harm or benefit those who they affect most directly.[36] This conclusion in turn renders it extremely difficult to provide an account of our obligations in relation to the use of these technologies.[37]

Unfortunately, the obvious alternative way of thinking about these dilemmas, as involving changes in the situation of a person identified independently of their particular genetic make-up, is equally, if not more, problematic. Baily offers this alternative in a passage immediately following that quoted above.

My picture is of a "disembodied soul," the essence of my yet-to-be-born child, waiting to be inserted into a baby shaped container, with me standing there to make sure my child's soul gets into a well functioning container.[38]

This way of thinking has the advantage of allowing us to talk of benefiting and harming particular individuals and therefore of grounding an obligation to use these technologies to promote the wellbeing of those people who are brought into existence. However, it has the obvious disadvantage of requiring an account of personal identity which is independent of any physical substrate. This is enough to render it unattractive to most contemporary thinkers.[39]

It is not just critics from the disability community, then, whose arguments run into trouble in the face of the notoriously difficult to set problems in metaphysics and the philosophy of personal identity which surround choices concerning who comes into existence. The possibility that I wish to explore in the rest of this chapter is that, rather than wait for these problems to be solved before attempting to engage with the disability critique, we might make some tentative progress if we approach the question of the identity of persons and expressive content of various choices and policies as *political* issues. That is, as issues which are, at least in part, about contestation for power between different groups of people.

Eugenics, 'old' and 'new'

One important reason for interpreting the expressivist critique as, at least in part, a political concern for relationships between social groups is that doing so draws our attention to the morally disturbing similarities between the sorting society and the eugenic programmes of the past.

There are, at least, three important similarities and one important dissimilarity between the 'new eugenics' of the sorting society and the 'old eugenics' of the 1930s.

The dissimilarity has been widely advertized and a number of writers have taken it to render the new eugenics fundamentally different from the old. Whereas the 'old' eugenics made use of the coercive power of the state to enforce its eugenic agenda and deprived individuals of their human rights in doing so, the 'new' eugenics is organized around the rhetoric of individual choice and respect for human rights, and abjures the use of the coercive powers of the state.[40] Given that the most obvious and dramatic atrocities associated with the old eugenics were for the most part carried out using the coercive powers of the state, this is an important difference.

However, although the new eugenics disavows the coercion and state planning of the old eugenics, it shares with it a deep-seated belief that genes matter. The project of the sorting society remains 'eugenic' – concerned with ensuring that children are of 'good birth'. In so far as it relies on technologies of genetic diagnosis to attempt to secure this goal, the sorting society requires that we can make plausible assumptions about the sorts of lives people are likely to have on the basis of their genes. It requires both that we can sort embryos and that we can distinguish between the different sorts of people they are likely to become.

It is also equally true of the new genetics as of the old that some sorts of people will not be chosen to be brought into the world once the use of technologies of genetic selection becomes widespread. Indeed, the effect of the widespread adoption of the new eugenics is likely to produce similar outcomes to those that were the goal of the old eugenics. It is embryos that might have become 'less than perfect' children that will be selected for termination or that will not be implanted. Children that might once have been born with a disability, or born female, or born with any of a range of character traits deemed to be socially undesirable, will cease to appear in the world. In their place will be born children who are strong and healthy, good-looking, male, and who have all the marks of local privilege.

The history of the use of these technologies to this date supports these predictions. In societies where daughters are considered less valuable than sons their use has led to fewer girls and more boys being born.[41] The range of conditions tested for has increased, as has the number of diagnoses which have been used to justify selecting another embryo. Moreover, arguments are beginning to appear in both the scientific and philosophical literature that we are obligated to use these technologies to produce children who will have the 'best possible' lives.[42] That is to say, that the distinction between therapeutic and non-therapeutic uses of these technologies is already being blurred in the philosophical literature. One suspects it is only a matter of time before wealthy couples begin using these technologies to ensure that their child has 'superior genes'.[43]

In part, this pattern of likely outcomes is a function of the pre-existing preferences of parents. That is, these technologies are likely to be taken up to prevent the birth of children with character traits that are currently socially disvalued and/or to ensure the birth of children with traits that are considered to be valuable. However, the existence of technologies of genetic selection also plays a significant role in *shaping* these preferences. Once these technologies become available it becomes more difficult for parents to refuse them, or to choose to make the 'less popular' choice and proceed with a pregnancy that may lead to the birth of a child with a disability, because the consequences of making various choices depend in part on the aggregation of the choices of others in similar circumstances. For instance, if other parents make use of a technology of genetic selection to prevent the birth of children with a particular condition then the availability of social and medical support services to support children with that condition will decrease as a result of demand for these services being reduced.[44] In this fashion, the technologies which make possible the sorting society condition their own use. Indeed, many critics have expressed concern that the existence of these technologies will establish a 'genetic rat race' which will effectively force people to adopt them or else risk that their children will be left behind in competition for scarce social goods.

However, it is more than the sum of the individual choices of parents, or even 'market forces', which leads to the use of these technologies to select in favour of certain sorts of people. The idea that it is rational to prefer some sorts of children to others is built into the very foundations of the sorting society. As Nancy Press[45] has remarked, it is the choices 'made available' which reveal the attitudes driving the new eugenics. While programmes of genetic screening and testing are usually established and defended on the basis of offering choice to parents, it is clear that they would not exist except for the belief that having such a choice is important *because* there are *good reasons* to prefer a child *without* a disability. That is, the justification for the dedication of resources to making such a choice possible relies implicitly on the notion that many people would wish to exercise such a choice in a particular direction. Governments do not, after all, typically fund research into screening technologies in order to allow people to choose the colour of their children's eyes, or hair colour, or to allow them to try to have children with particularly shapely elbows. This suggests that while the coercive powers of the state are not mobilized by the new eugenics, other aspects of the state are. The state which rules the sorting society is not 'neutral' on the question of what sorts of people should be born. In providing funding for the development of some tests and not others, and in funding public access to a limited range of tests, the state necessarily reveals beliefs about which choices are worth having, and in turn which conditions one would be justified in choosing against.[46] In this way, the new eugenics will still involve state policy about what sorts of people there should be.

Indeed, such policy is often explicit – or, at least, easily discernible – in the justifications given in various bureaucratic debates about whether to support routine genetic screening for particular conditions. When costing various programmes, government agencies typically compare the amount of money it will cost the government to make a screening technology available to a certain cohort of parents with the amount of money which would be saved by eliminating the need to provide medical care and social services to those people who might otherwise be born with the condition the test is designed to diagnose.[47] If the state program will save more money than it will cost then the argument that it should be funded becomes straightforward. The justification of the policy therefore essentially refers to how many births/lives it will prevent.

An important lesson to be drawn from a political concern with power and structure about the sorting society, then, is that the expressivist critique may be especially strong when it is directed towards applications of technologies of genetic selection that are socially sanctioned and state funded.[48]

A much criticized feature of the expressivist critique is that it sometimes seems to rely on an assumption about the motivations of individual parents, when they decide to terminate a pregnancy on the basis of a diagnosis of a possible disability, or to choose one embryo over another on the same basis.[49] That is, in some formulations of this critique, it appears as though the sentiment the parents' actions express is the attitude that they (must) have towards persons with disabilities.[50] However, as I observed earlier, parents making decisions about which embryo to choose may not be thinking about the worth of the lives of anyone at all.[51] Instead, they may simply be trying to secure for their children certain ordinary human capacities, the absence of which does not render a life not worth living but can make achieving various social and economic goods much more difficult. In so far as they think about the 'worth' of various lives at all – perhaps because they have been exposed to the expressivist critique – they may, in fact, affirm the equal worth of all human lives. Thus, if the expressivist critique relies upon facts about the actual intentions of parents making these sorts of decisions, it may apply only rarely.[52] Furthermore, the decisions of individual couples have only an infinitesimal effect on the relations that exist between those social groups whose members might claim to have an interest in their decisions. As a result, it is difficult to see how either the expressive content or the consequences of their decisions could justify the interference with the liberty of the parents which would be involved in preventing them from acting on their decision.

However, the situation is quite different when we are considering the expressive content of government policies. These policies often reveal motivations and intentions which are formed only fleetingly, if at all, in the course of the deliberations of individual parents. As I argued above, even if the choice as to how to act on a test result is left up to prospective parents, the justification of the policy that makes it available refers to the expectation that this choice will often be made in a certain way. Moreover, these policies will have a dramatic impact on relations between people with disabilities and people without disabilities. While the choices of individuals will have only a marginal effect on what sort of people will exist in the future, the consequences of adopting various policies about researching, funding, or introducing screening programmes may very well include ensuring that certain sorts of people will disappear from the community over the course of a few generations. The argument that various applications of technologies of genetic selection express something is therefore much stronger when it is policies rather than individual choices which are being considered for their expressive content.

This last observation draws our attention to a third similarity between the old and the new eugenics, which is less often remarked upon, which is that the new eugenics, like the old, overwhelmingly involves one sort of people making decisions which primarily affect another. The old eugenics involved those who believed themselves to be genetically superior making decisions about those they believed to be 'unfit'. In the new eugenics, it is, by and large, persons without disabilities who make decisions about funding research and programmes of genetic screening, whereas it is persons with disabilities who feel threatened by them and whose communities may well disappear as a result of them.

The significance of this fact for debates about technologies of genetic selection is hinted at by an otherwise puzzling feature of the expressivist critique. One of the many striking things about the expressivist critique is how weak the conclusion of the argument is, given its apparent ambition initially and also the strength of the claims made in the course of the argument. If it really were the case that those making the expressivist critique believed that the use of technologies of genetic selection to select against the birth of persons with disabilities was motivated by, and expressed, a profound lack of respect for a significant portion of the population, one might expect them to go on to mount an argument against the use of these technologies in this fashion altogether. Instead, the usual conclusion of the 'expressivist critique' is only that parents should be exposed to the opinions and experiences of those with the disability they are concerned about, or those caring for them, before making their decision.[53] That is, what is ultimately contested in this argument is the process whereby these decisions are made, and who makes them, rather than the content of the decisions. The negative attitudes towards people with disabilities that the use of technologies of genetic selection expresses are expressed not so much by the decisions that are actually made but by the fact that people with disabilities are not included in the decision-making about their use. It is the relation between groups of people within a generation, between people with disabilities and people without disabilities, that is crucial here, rather than the relationship between parents and their (possible) future children.

The expressivist critique revisited

This way of understanding what is at stake in debates about technologies of genetic selection cannot avoid the problematic implication, noted above, that people with disabilities are, apparently, according to this way of thinking, essentially constituted by their disabilities. Why should people with disabilities or – perhaps the more politically urgent question – people without disabilities care about what happens to a social group defined by disability unless membership of this group marks something deep and important about those within it? In many other policy debates people with disabilities are often insistent that they should *not* be characterized as 'disabled persons' and that a properly inclusive society would *not* single out people with disabilities for special treatment.[54] The apparent tension between these claims can be at least partially reconciled if we recognize them both as being understandable responses to a political dilemma that confronts members of marginalized social groups across a broader range of issues. In order to see how this might be the case, it will help to first consider an analogy that is often referred to in the course of the expressivist critique between selecting against disability and selecting against race or sex.[55]

I noted above that sorting embryos according to their sex or 'race' has far fewer defenders than selection on the basis of possible disability. As critics from the disability community have rightly pointed out, if the technologies of the sorting society were used primarily to select on the basis of skin colour, for instance, we would think that this is likely both to reflect and to express racism.[56] In particular, any state-funded screening programme in Europe, the United States, the United Kingdom, or Australia designed to allow parents the option of terminating pregnancies likely to lead to the birth of dark-skinned children would rightly be condemned as racist. Similarly, it is clearly the case that the use of sex selection technologies to select against the birth of female children in India, Bangladesh and China is a sexist practice.[57] This remains true even if the intentions of a particular parent are not the product of gender hierarchy.

There are, I think, at least two good reasons why our intuitions should be so much clearer in these cases.

Firstly, it is much more widely recognized in relation to racism and sexism that what a person's actions 'express' cannot be entirely determined by them nor settled by referring to their intentions.[58] Instead, because meaning is social, the expressive content of our utterances and our actions is determined by what other people will understand of them. This means that our utterances and actions may even have expressive content contrary to that which we intend. Genetic selection on the basis of race (or sex) is racist (or sexist) because it is continuous with and perpetuates historical patterns of oppression on the basis of race (or sex) even where the intentions of parents are beneficent. The social context of these decisions over-determines their content.[59]

Secondly, the doubt canvassed above, about whether such uses of these technologies are directed towards and affect a certain 'sort' of person, does not arise here. Race and sex are social categories which structure contemporary societies to such an extent that we cannot help but acknowledge the different sorts of people which exist either side of the divisions they establish.

Thus what makes the idea that selection on the basis of sex or race is sorting persons more immediately obvious than the case of selection on the basis of disability is that the majority of the community is more committed to these categories as describing socially important phenomena. (This is also why 'disability' in turn marks out a – more or less – distinct social group in a way that being the youngest child in the family does not.[60]) Moreover, because of the prevalence and influence of 'gene talk' in modern societies, in part as a consequence of the development of technologies of genetic selection, these social identities include a genetic component. That is, we imagine them to be 'in the genes'.[61] As a result, sorting *embryos* on the basis of sex or (putative) race can also be understood as distinguishing between different sorts of persons. Because the social practices and institutions organized around the categories of race and sex have real social force, changes in the relations between these groups are likely to be important to those within them. Selecting future persons on the basis of race or sex clearly will have such an effect and is therefore politically and ethically controversial.

This remains true even if, in more philosophical and reflective moments, we acknowledge that these distinctions are often arbitrary and morally pernicious. Morally pernicious systems of discrimination can have real social effects which extend over long periods of history and across a wide range of social institutions. These systems both rely upon and produce discriminatory social categories, such as race, caste and social class. These categories become categories of social organization with very real effects on the shape of society and the lives of those within them. As a result they become terms with genuine reference to classes of people. Yet at another level they remain founded on fictions.

Social policy – and criticism – in relation to inequalities of this sort confronts a dilemma. In order to identify and remedy those injustices that we wish to address we must formulate policy in terms of these categories.[62] However, doing so strengthens the very categories which constitute and sustain these unjust social relations. If, alternatively, we try to prefigure the future of just social relations between persons that we desire, and formulate policy in terms which do not presume these categories, then there is a very real risk that we will be unable to address existing injustices.

The dilemma facing those who would put forward the expressivist critique of the use of technologies of genetic selection to prevent the birth of persons with disabilities is therefore an instance of a dilemma facing members of oppressed and socially marginalized groups more generally: how to negotiate the troubled waters of identity politics in a society shaped by existing and historical injustices? This dilemma perhaps presents itself with extra urgency in the debate about selection on the basis of *disability* because this category appears, at least at times, on both sides of the debate as primarily 'genetic'. Both critics and advocates of technologies of genetic selection are often committed to the implication that some genetic conditions are constitutive; they determine what 'sort' of person someone is. Equally well, both critics and advocates have strong reasons to deny that the identity of a future person is coded in the genes of each embryo. In some circumstances then both sides of these debates imply that our genes make us who we are, whereas in others they are quick to insist that it is environment and society that determine what sort of person embryos become.

The unresolved tension in these discussions of the relation between genetics and identity, should, I believe, be understood as reflecting this same dilemma. The question 'How much are we constituted by our genes?' is – at least when we are discussing the ethics of the use of genetic technologies – as much a question of politics and sociology as it is a question of biology, ontogeny, psychology or metaphysics. It is social understandings that determine which conditions are contingent to, and which are constitutive of, personal identity. What sorts of social identity have a genetic component is itself a matter of political contestation. Genetic technologies and the debate surrounding them therefore play a central role in determining what the relationship is between genes and persons. This means that both critics and advocates of technologies of genetic selection have reason to be more careful about the implications of their arguments for the relationship between genes and identity and therefore for the force of the expressivist objection. The contentious analogy between sorting embryos on the basis of disability and on the basis of sex or race may have more to teach us, then, about the ethics of the sorting society than simply lending intuitive support to the expressivist critique. It may be that further insights into, and progress in, the debate about the sorting society could be achieved by looking more closely at debates about multicultural and identity politics.[63] However, this project would be a larger task than I could hope to achieve here.

Conclusion

I must acknowledge that my discussion in this chapter has stopped short of an evaluation of the expressivist critique. I have tried to show how this critique is more challenging if understood as a critique of existing relations between social groups. I have also tried to show how the process of contesting these relations is itself implicated in the construction of a social identity around disability which itself is part of what is being contested.

However, I have not tried to settle the ethics or the politics of the construction of such an identity. Constructing a social identity and community is not without its costs, especially when such construction occurs around an already existing identity that has been, for the most part, shaped by the responses of others and which has, in many circumstances, appeared largely as an obstacle to respect for the rights of persons with disabilities. However, in the first instance at least, judgements about the consequences and relative merits of various political strategies in relation to identity are best made by those who are more likely to find themselves identifying or being identified with the group whose members will be most affected. For this reason, as a person who does not identify as a person with a disability, I have hesitated to enter into a debate about whether people with disabilities might be better off denying or affirming this label.

Nor have I tried to determine whether the expressivist critique succeeds in establishing an all-things-considered objection to the use of technologies of genetic selection to select against the birth of persons with disabilities. Key questions here will obviously be whether a concern for relations between social groups can justify the restriction of individual liberty imposed by preventing prospective parents from using these technologies as they wish; and how any such concern weighs in relation to any duties of Procreative Beneficence – to promote the wellbeing of those we bring into the world – that we acknowledge.[64] However, in so far as I have resiled from evaluating the force of arguments derived from a concern for relations between particular social groups I am equally unable to resolve these further questions. Moreover, the interpretation of the expressivist critique I have been advocating suggests that the answer to these questions is sensitive to social context. That is, the force of the expressivist critique will depend on facts about the extent to which people identify themselves and others as being certain 'sorts' of people in the societies which are considering the ethics of genetic selection.[65]

Nonetheless, I hope that the interpretation of the expressivist critique I have offered here may usefully contribute to the larger project of evaluating the critique. Unless philosophers endeavour first to understand why this critique seems compelling to those who advance it, we are unlikely to convince them that we have considered it adequately. My examination of an influential objection to the expressivist critique, based on a series of analogies between ordinary therapeutic and selective genetic interventions, has highlighted the way both critics and proponents of the use of technologies of genetic selection must negotiate a difficult set of dilemmas surrounding the relationship between genes and identity. Paying attention to the political dimensions of this negotiation lends strength to the expressivist critique in three ways. Firstly, it highlights important continuities and similarities between the 'old' and the 'new' eugenics, in particular that the state which governs the modern sorting society is not neutral about the sorts of people who will be born in the future. Secondly, it draws our attention to the fact that the policies which the state adopts concerning the use of technologies of genetic selection express attitudes towards persons with disabilities which plausibly may be morally evaluated. Thirdly, it shows how the expressive content of these policies may be as much a function of the relations between the people who determine the policies as it is about their content. These lessons suggest that the expressivist critique may have more force and substance than is often appreciated.

Finally, acknowledging the crucial role of claims about identity in debates about the ethics of the sorting society allows us to acknowledge the high degree of reflexivity in the expressivist critique and in responses to it. Contributions to these debates inevitably help constitute our understandings of the relation between genetics and identity at the same time as they contest these understandings. This suggests that there may yet remain lessons to be learnt about the ethics of the sorting society from other debates about relationships between different sorts of people.[66]

ACKNOWLEDGEMENTS

I would like to thank Justin Oakley, Jacqui Broad, Debra Dudek, Toby Handfield and Emilio Mora for helpful comments and discussion over the course of the development of this paper.

NOTES AND REFERENCES

1. The most powerful of these technologies, pre-implantation genetic diagnosis, remains a comparatively rare phenomenon largely because it requires conception to occur via in-vitro fertilization. However, various forms of prenatal screening, with the option of termination, are now a routine part of antenatal care in much of the industrialized world.
2. A. Asch, 'Reproductive technology and disability' in S. Cohen and N. Taub (eds.), *Reproductive Laws for the 1990s* (Clifton, NJ: Humana Press, 1988) 69–124; A. Asch, 'Why I haven't changed my mind about prenatal diagnosis: reflections and refinements' in E. Parens and A. Asch (eds.), *Prenatal Testing and Disability Rights* (Washington, DC: Georgetown University Press, 2000) 234–58.
3. S. Wendell, *The Rejected Body* (New York: Routledge, 1996).
4. M. Saxton, 'Disability rights and selective abortion' in R. Solinger (ed.), *Abortion Wars: A Half Century of Struggle* (Berkeley and Los Angeles: University of California Press, 1997) 374–95.
5. D. Kaplan, 'Prenatal screening and its impact on persons with disabilities' (1993) 36(3) *Clinical Obstetrics and Gynecology* 605–12.
6. A. Buchanan, 'Choosing who will be disabled: genetic intervention and the morality of inclusion' (1996) 13(1) *Social Philosophy and Policy* 18–46; E. Parens and A. Asch (eds.), *Prenatal Testing and Disability Rights* (Washington, DC: Georgetown University Press, 2000).
7. T. Stainton, 'Identity, difference and the ethical politics of prenatal testing' (2003) 47(7) *Journal of Intellectual Disability Research* 533–9.

8. A. Buchanan, D. W. Brock, N. Daniels and D. Wikler, *From Chance to Choice* (Cambridge: Cambridge University Press, 2000) 265.
   In what follows I will, for the most part, restrict my discussion to the use of the technology of pre-implantation genetic diagnosis to this end in order to avoid the complexities arising from the psychological and physical trauma to the mother which may result from terminating a pregnancy.

9. A. Davis, 'Women with disabilities: abortion and liberation' (1987) 2(3) *Disability, Handicap & Society* 275–84; D. Kaplan, 'Prenatal screening and its impact on persons with disabilities', note 5 above, 605–12; S. Wendell, *The Rejected Body*, note 3 above, 83, 153–154; M. Saxton, 'Disability rights and selective abortion', note 4 above, 374–95.

10. A. Buchanan, D. W. Brock, N. Daniels and D. Wikler, *From Chance to Choice*, note 8 above, 265.

11. A. Buchanan, D. W. Brock, N. Daniels and D. Wikler, *From Chance to Choice*, note 8 above, 276–81; B. Steinbock, 'Disability, prenatal testing, and selective abortion' in E. Parens and A. Asch (eds.) *Prenatal Testing and Disability Rights* (Washington, DC: Georgetown University Press, 2000) 108–23.

12. N. Agar, *Liberal Eugenics: In Defence of Human Enhancement* (Oxford: Blackwell Publishing, 2004); J. L. Nelson, 'Prenatal diagnosis, personal identity, and disability' (2000) 10(3 (September)) *Kennedy Institute of Ethics Journal* 213–28; J. Savulescu, 'Procreative beneficence: why we should select the best children' (2001) 15(5) *Bioethics* 413–26.

13. A. Buchanan, D. W. Brock, N. Daniels and D. Wikler, *From Chance to Choice*, note 8 above, 267–76; J. L. Nelson, 'Prenatal diagnosis, personal identity, and disability', note 12 above, 213–28; B. Steinbock, 'Disability, prenatal testing, and selective abortion', note 11 above, 108–23.

14. A. Buchanan, D. W. Brock, N. Daniels and D. Wikler, *From Chance to Choice*, note 8 above, 275; J. L. Nelson, 'Prenatal diagnosis, personal identity, and disability', note 12 above, 213–28.

15. J. Harris, *Clones, Genes and Immortality: Ethics and the Genetic Revolution.* (Oxford: Oxford University Press, 1998) Chapter 3.

16. N. J. Zohar, 'Prospects for "genetic therapy" – can a person benefit from being altered?' (1991) 5(4) *Bioethics* 275–88.

17. A. Buchanan, D. W. Brock, N. Daniels and D. Wikler, *From Chance to Choice*, note 8 above, Chapter 6; J. L. Nelson, 'Prenatal diagnosis, personal identity, and disability', note 12 above, 213–28; J. Savulescu, 'Procreative beneficence: why we should select the best children', note 12 above, 413–26.
   Moreover, while I have sketched this argument in the context of the therapeutic intervention, a similar form of argument is possible in relation to the moral permissibility of – and perhaps even obligation to pursue – enhancements. That is, if we are willing to concede that it is morally permissible, or perhaps admirable, or perhaps even obligatory to provide a child with various benefits through a non-therapeutic environmental intervention such as vaccination, cosmetic surgery, or character-shaping education then, if we can similarly imagine that this result might be able to be achieved by intervention in utero, somatic cell gene therapy, germ line gene therapy, or PGD and selection amongst embryos, it seems that we

should also hold that these interventions would also be morally permissible, admirable, or perhaps obligatory (Savulescu, 2001). To discuss the issues raised by the use of technologies of genetic selection to enhance human beings would take me too far from the main task of my argument, which is to interrogate the competing accounts of identity that underlie debates about their use to prevent disability. However, in so far as such accounts are also crucial to debates about human enhancement, I hope the discussion which follows will also illuminate those debates.

18. A. Asch, 'Why I haven't changed my mind about prenatal diagnosis: reflections and refinements', note 2 above; J. L. Nelson, 'Prenatal diagnosis, personal identity, and disability', note 12 above, 220.

19. A. Asch, 'Why I haven't changed my mind about prenatal diagnosis: reflections and refinements', note 2 above, 234–58.

20. A. Buchanan, D. W. Brock, N. Daniels and D. Wikler, *From Chance to Choice*, note 8 above, 264–74.

21. T. Stainton, 'Identity, difference and the ethical politics of prenatal testing', note 7 above, 534.

22. J. L. Nelson, 'Prenatal diagnosis, personal identity, and disability', note 12 above, 219.

23. B. K. Rothman, *Genetic Maps and Human Imaginations: The Limits of Science in Understanding Who We Are* (New York: Norton & Co., 1998).

    This is not to say that, in relation to their causal role in producing any given variation in phenotype, genetics and the environment are equally important, only that without specifying both we cannot determine the phenotype of the organism at all.

24. J. L. Nelson, 'Prenatal diagnosis, personal identity, and disability', note 12 above, 219; T. Stainton, 'Identity, difference and the ethical politics of prenatal testing', note 7 above, 534.

25. A. Buchanan, 'Choosing who will be disabled: genetic intervention and the morality of inclusion', note 6 above, 18–46; J. L. Nelson, 'The meaning of the act: reflections on the expressive force of reproductive decision making and policies' (1998) 8(2) *Kennedy Institute of Ethics Journal* 165–82; J. L. Nelson, 'Prenatal diagnosis, personal identity, and disability', note 12 above, 213–28.

26. N. Press, 'Assessing the expressive character of prenatal testing: the choices made or the choices made available?' in E. Parens and A. Asch (eds.), *Prenatal Testing and Disability Rights* (Washington, DC: Georgetown University Press, 2000) 215.

27. J. L. Nelson, 'Prenatal diagnosis, personal identity, and disability', note 12 above, 219.

28. M. A. Baily, 'Why I had amniocentesis' in E. Parens and A. Asch (eds.) *Prenatal Testing and Disability Rights*. (Washington, DC: Georgetown University Press, 2000) 66.

29. N. Press, 'Assessing the expressive character of prenatal testing: the choices made or the choices made available?', note 26 above, 215.

30. P. Kitcher, *The Lives to Come: The Genetic Revolution and Human Possibilities* (New York: Simon and Schuster, 1996) Chapter 11.

31. B. K. Rothman, *Genetic Maps and Human Imaginations: The Limits of Science in Understanding Who We Are*, note 23 above.

32. Moreover, in other debates about bioethics, especially about new reproductive technologies, a good deal of weight is placed on genetic identity – especially genetic relatedness. In many moral – and legal – contexts genes determine whose children are whose; decisions about which embryos to implant, or bring to term, are regularly made on the basis that their genes determine key aspects of their identity: K. D. Alpern, 'Genetic puzzles and stork stories' in K. D. Alpern (ed.), *The Ethics of Reproductive Technology* (Oxford: Oxford University Press, 1992) 147–69.

33. B. Jennings, 'Technology and the genetic imaginary: prenatal testing and the construction of disability' in E. Parens and A. Asch (eds.), *Prenatal Testing and Disability Rights* (Washington, DC: Georgetown University Press, 2000) 124–44.

34. D. Parfit, *Reasons and Persons* (Oxford: Clarendon Press, 1984).

35. J. Glover, 'Future people, disability, and screening' in J. Harris (ed.), *Bioethics* (Oxford: Oxford University Press, 2001), 429–44.

    An important exception here is where decisions about genetic selection might lead to the birth of children with lives so filled with pain and suffering and devoid of potentially rewarding experiences as to be 'not worth living'. In situations where it would be rational for the person who comes into existence to prefer to be dead, it seems that we *can* speak of harming someone in the act of bringing them into existence.

36. D. Brock, 'The non identity problem and genetic harms – the case of wrongful handicaps' (1995) 9(3/4) *Bioethics* 269–75.

    This way of understanding personal identity also appears to have strange implications for the ethics of our treatment of embryos. It looks as though on this account it is possible to benefit or harm embryos (and the individuals they will develop into) in morally significant ways but that it is not morally wrong to kill them. Actions which affect embryos which fall short of destroying the embryo must be evaluated in relation to the impacts they might have on the future well-being of the person this embryo will become. That is, the baseline for determining harm or benefit is the interests of the person who would have developed from an embryo if it retained the chromosomes it had at conception. However, actions which destroy an embryo and which therefore lead to another person being born, do not harm anyone. It seems, then, that we could inflict multiple successive harms on an embryo over an extended period with each action being morally wrong, except for the last one, which renders the embryo non-viable. That the last of a series of identical harmful actions should fail to be harmful because it proves fatal to the embryo is counter-intuitive to say the least.

37. D. Brock, 'The non identity problem and genetic harms – the case of wrongful handicaps', note 36 above, 269–75; C. B. Cohen, ' "Give me children or I shall die!" New reproductive technologies and harm to children' (1996) 26(2) *Hastings Center Report* 19–27; J. Feinberg, 'Wrongful life and the counterfactual element in harming' (1987) 4(1) *Social Philosophy and Policy* 145–78; Jonathon Glover, 'Future people, disability, and screening', note 35 above, 429–44; J. McMahan, 'Wrongful life: paradoxes in the morality of causing people to exist' in John Harris (ed.), *Bioethics* (Oxford: Oxford University Press, 2001) 445–75; B. Steinbock and R. McClamrock, 'When is birth unfair to the child?' (1994) 24 *Hastings Center*

Report 15–21; C. Strong, 'Harming by conceiving: a review of misconceptions and a new analysis' (2005) 30 *Journal of Medicine and Philosophy* 491–516.

38. M. A. Baily, 'Why I had amniocentesis', note 28 above, 66.

39. Although a recent paper by Jeffrey Reiman in *Philosophy & Public Affairs* does seem to embrace this alternative: J. Reiman, 'Being fair to future people: the non-identity problem in the original position' (2007) 35(1) *Philosophy & Public Affairs* 69–92.

40. N. Agar, *Liberal Eugenics: In Defence of Human Enhancement*, note 12 above, 5–6; A. Buchanan, D. W. Brock, N. Daniels and D. Wikler, *From Chance to Choice* note 8 above, 55–60; J. Savulescu, 'Procreative beneficence: why we should select the best children', note 12 above, 413–26; L. M. Silver, *Remaking Eden: Cloning, Genetic Engineering and the Future of Human Kind* (London: Pheonix, 1999) 255; D. Wikler, 'Can we learn from eugenics?' (1999) 25(2) *Journal of Medical Ethics* 183–94.

41. N. Oomman and B. R. Ganatra, 'Sex selection: the systematic elimination of girls' (2002) 10(19) *Reproductive Health Matters* 184–8.

42. J. Savulescu, 'Procreative beneficience: why we should select the best children', note 12 above, 413–26.

43. L. M. Silver, *Remaking Eden: Cloning, Genetic Engineering and the Future of Human Kind*, note 42 above.

44. A. Asch and G. Geller, 'Feminism, bioethics, and genetics' in S. Wolf (ed.), *Feminism and Bioethics: Beyond Reproduction* (New York; Oxford: Oxford University Press, 1996) 325.

45. Nancy Press, 'Assessing the expressive character of prenatal testing: the choices made or the choices made available?', note 26 above, 214–33.

46. Bruce Jennings, 'Technology and the genetic imaginary: prenatal testing and the construction of disability', note 35 above, 124–44; Nancy Press, 'Assessing the expressive character of prenatal testing: the choices made or the choices made available?', note 26 above, 231.

47. Nancy Press, 'Assessing the expressive character of prenatal testing: the choices made or the choices made available?', note 26 above, 221–3.

48. *Ibid.*, 214–33.

49. J. L. Nelson, 'Prenatal diagnosis, personal identity, and disability', note 12 above, 213–28.

50. A. Buchanan, 'Choosing who will be disabled: genetic intervention and the morality of inclusion', note 6 above, 18–46.

51. B. Steinbock, 'Disability, prenatal testing, and selective abortion', note 11 above, 108–23.

52. A. Buchanan, 'Choosing who will be disabled: genetic intervention and the morality of inclusion', note 6 above, 18–46.

53. A. Asch, 'Reproductive technology and disability', note 2 above, 89–92; D. Kaplan, 'Prenatal screening and its impact on persons with disabilities', note 5 above, 605–12.

54. A. Asch, 'Reproductive technology and disability', note 2 above, 69–124.

55. *Ibid.*, 85–6; S. Wendell, *The Rejected Body*, note 3 above, 81.

56. A. Asch, 'Reproductive technology and disability', note 2 above, 85–6; S. Wendell, *The Rejected Body*, note 3 above, 81.

57. N. Oomman and B. R. Ganatra, 'Sex selection: the systematic elimination of girls', note 41 above, 184–8.

58. J. L. Nelson, 'Prenatal diagnosis, personal identity, and disability', note 12 above, 213–28.

59. This remains true even if it is also true that people born with dark skin face substantial disadvantages in the society in which the decision takes place. More controversially, I believe it may be true even if a concern for skin colour was motivated by a purely 'medical' concern for the connection between skin pigmentation, exposure to UV radiation in sunlight and vitamin D deficiency. The history of racism makes any use of medical technology to alter the composition of racial groups extremely problematic. Resistance to this conclusion stems largely, I believe, from the idea that racism is primarily a matter of intention. My contention here is precisely that actions, especially actions by institutions, can have a political character, which is not a function of the intentions behind them but rather of their nature and consequences viewed in a historical context.

60. A. Asch, 'Why I haven't changed my mind about prenatal diagnosis: reflections and refinements', note 2 above, 234–58; S. Wendell, *The Rejected Body*, note 3 above; T. Stainton, 'Identity, difference and the ethical politics of prenatal testing', note 7 above, 533–9.

61. We can intuitively understand the idea of selecting on the basis of race even though race is not a category with any real genetic (indeed, scientific) basis. While traits which are associated with racial stereotypes, such as skin, eye and hair colour and facial bone structure, undoubtedly have a genetic basis, 'race' – a social category – does not.

62. M. Minow, *Making All the Difference: Inclusion, Exclusion and American Law* (Ithaca: Cornell University Press, 1990); S. Wendell, *The Rejected Body*, note 3 above; I. M. Young, *Justice and the Politics of Difference*, (Princeton, New Jersey: Princeton University Press, 1990).

63. T. Stainton, 'Identity, difference and the ethical politics of prenatal testing', note 7 above, 533–9; S. Wendell, *The Rejected Body*, note 3 above; I. M. Young, *Justice and the Politics of Difference*, note 62 above.

64. It seems highly unlikely, for instance, that such concern should prevent us from making use of screening technologies for the purpose of preventing the birth of children who would have 'a life not worth living' such that it would be rational for them to prefer to be dead.

65. T. Stainton, 'Identity, difference and the ethical politics of prenatal testing', note 7 above, 537.

66. W. E. Connolly, *Identity/Difference: Democratic Negotiations of Political Paradox* (Ithaca: Cornell University Press, 1991); M. Minow, *Making All The Difference: Inclusion, Exclusion and American Law*, note 62 above; T. Stainton, 'Identity, difference and the ethical politics of prenatal testing', note 7 above, 533–9; I. M. Young, *Justice and the Politics of Difference*, note 62 above.

10

Overstating the biological: geneticism and essentialism in social cloning and social sex selection

Mianna Lotz

Introduction

Technological expansion of our capacity directly to influence reproductive outcomes has, over the past four decades or thereabouts, significantly extended our scope for making what we are here referring to as 'sorting' decisions: reproductive decisions about the kinds of people to be born. A central concern of this book is with evaluating the preferences and values that govern the sorting decisions we make in the sphere of assisted procreation. Critical scrutiny of reproductive motivations is not always welcomed, due in no small part to the modern philosophical commitment to a liberal view of reproduction as falling within the private sphere of human life and decision-making. However, acceptance of the liberal view does not preclude an examination of the moral basis of procreative preferences. In so far as we seek to lead so-called 'examined' and autonomous lives, in which our significant motivations and preferences are ones that we are able to endorse and act upon, it is entirely appropriate that we be willing to hold ourselves to account for the preferences that influence our procreative decisions.

To that end, my aim here is to address notable critical 'blind-spots' in regards to two particular procreative preferences: the preference for biological relatedness, sometimes offered as an argument in favour of cloning for non-therapeutic, purely reproductive purposes (herein referred to as 'social cloning'); and the preference for a 'balanced family', sometimes offered in support of non-medical sex selection (herein termed 'social sex selection'). In recent discussions there has been a tendency towards uncritical acceptance of both of these as reasons capable of justifying even quite extreme reproductive interventions like human cloning. I want to suggest, however, that both

The Sorting Society: The Ethics of Genetic Screening and Therapy, ed. Loane Skene and Janna Thompson. Published by Cambridge University Press. © Cambridge University Press 2008.

preferences warrant closer critical scrutiny than they typically receive, and my goal here is to contribute towards overcoming that deficiency.

Importantly, in examining the extent to which these preferences have the moral warrant they are assumed to have, I am not thereby committing myself to any particular substantive view about the permissibility of the practices they are intended to justify. Nor need I. If I succeed in showing that these preferences are not morally innocuous to the extent typically assumed, it remains a distinct and here-unaddressed question precisely what ought to be the practical implications of that evaluation. We may, after all, quite reasonably protect decision-making freedoms and permit even those practices that we regard to be morally problematic, on the grounds that their proscription would be more morally problematic still. Accordingly, I will not here comment on whether social cloning or social sex selection should be withheld from those motivated by the preferences discussed here. My discussion deals exclusively with the prior question of whether the procreative preferences in question, and the arguments underpinning them, are morally innocuous.

Geneticist assumptions in social cloning: the 'psychic similarity' and 'self-knowledge' arguments

It has been argued that the strongest moral argument in favour of social cloning is that it will allow the otherwise infertile to have children to whom they are genetically related.[1] A firm yet widely unquestioned belief in the moral significance and value of a close genetic connection between parents and children lies at the heart of that argument. The question is whether such a belief is itself well supported. Genetic relatedness is widely valued, but to avoid conflating empirical with moral claims, the moral significance of genetic relatedness requires justification beyond empirical observations concerning actual preferences.

Various attempts have been made to account for the putative moral value of parent–child genetic relatedness. Amongst them are arguments that genetic connection ensures better care for offspring, or results in more satisfying parenting and greater parent–child intimacy.[2] These arguments have been considered elsewhere, and I will not pursue them further here. The particular arguments I want to examine are, firstly, that genetic relatedness is valuable because it secures a morally desirable parent–child psychological similarity; and secondly, that genetic relatedness is valuable as a source of self-knowledge.

Brenda Almond and Michael Tooley are amongst those who embrace the 'psychic similarity' view, granting biology fairly extensive powers in the determination of a broad set of psychological traits and characteristics. Almond includes in this set '... shared attitudes, appraisals, interests, tendencies, common qualities of character, a common *Weltanschauung* – a characteristic way of looking at the world'.[3] Although she acknowledges that such psychic similarity cannot be assumed to exist between biological kin, she endorses the view that a complete absence of such psychic similarities and common sympathies is 'extraordinary', and she accords biological connectedness the primary explanatory role where they exist. Tooley's defence of the psychic similarity view is more extensive and tied to a normative argument that children benefit as a result of parents being better able to appreciate their point of view when they share a biological connection.[4] Granted, Tooley is careful to point out that the production of desired traits and abilities in one's offspring will be a function not merely of heredity but also of environment. Nevertheless he says:

> . . . given the greater psychological similarity that will exist between the child and one of his parents in such a case [of cloning], the relevant parent will better be able, at any point, to appreciate how things look from the child's point of view. So it would seem that there is a good chance both that such a couple will find childbearing a more rewarding experience, and that the child will have a happier childhood through [being] better understood.[5]

Tooley's view embodies both an empirical and a moral claim, the basis of each of which is questionable. Empirically, the suggestion that cloning would secure parent–child psychological similarity is surely misguided. It has not been conclusively established that genes are the basis of all of our psychological traits, least of all of the kinds of traits relevant to Tooley's (and Almond's) arguments.[6] Recall that for them, the traits of interest are those that contribute to a person's 'point of view'. The notion of a 'point of view' is of course hopelessly vague, and the range of traits that contribute to it very difficult to delineate. More specifically, though, Tooley mentions traits such as intelligence, persistence, determination and confidence in one's own abilities. The problem is that even if it is discovered that traits such as these do indeed have a genetic underpinning, such a discovery would not support an assumption that possession of the genetic basis for a psychological trait would guarantee or reliably predict for the realization or expression of that trait. As is now widely appreciated, gene expression is far too 'interactionist' to permit such an assumption: the vast majority of traits and conditions are polygenic, and genetic expression is multifactorial and dependent upon a complex interplay of the relevant genetic, proteome and socio-environmental conditions. While cloning might replicate genes, it cannot replicate the interactions of genes with other genes, or with proteins, or with socio-environmental conditions, all of which interactions play a decisive role in determining whether and which genes are expressed. Acknowledging this requires acknowledging, contra Almond and Tooley, that cloning can deliver no guaranteed nor even particularly reliable means for achieving parent–child psychological similarity.

Moreover, non-biological explanations can readily account for parent–child psychological similarity, where it occurs, without need for reference to biogenetic or heritability explanations. In the vast majority of cases of the sort that interest us here, psychological trait correlations are explicable by reference to context, socialization and learning – in other words, in exclusively social and environmental terms. This is not to claim that biogenetic factors play no role whatsoever, but rather that a presumption in favour of an exclusively biological explanation for parent–child similarity is unwarranted.

Granted, a proponent of the 'psychological similarity' argument might concede that socio-environmental factors play a significant role in the development of psychological traits, yet nevertheless insist that the value of psychological similarity is sufficient to justify attempts to *supplement* acknowledged socio-environmental causes with biological ones, precisely so as to maximize the likelihood of achieving parent–child psychological similarity. Assessing this possible rejoinder brings me to the second line of objection to the psychological similarity argument for social cloning. This line of objection does not question the argument's empirical claims but rather its normative claim, namely that parent–child psychological similarity is desirable and morally valuable. Even if we were to accept that genetic relatedness secures psychological trait similarity, should we accept that parent–child psychological similarity is a moral good, sufficient to justify reproductive interventions like social cloning as a means for achieving it? Are parental preferences for psychologically similar children morally innocuous?

On the contrary, I think we have good reason to doubt that it is necessarily in a child's interests to have parents whose desire for psychological similarity between themselves and their child is sufficiently strong to motivate them to undertake cloning in order to achieve it. Parental aspirations of this kind threaten a child's interests in being oriented to as a unique individual whose psycho-emotional characteristics are in important respects *developing* as well as distinct. We should not assume that children will be better off for the making of presumptions and prejudgements about their 'true' – because genetically given – inclinations, dispositions and nature. Nor, for that matter, can it be assumed that children necessarily benefit from *actualized* parent–child psychological similarity. What *is* plausibly regarded to be a benefit to a child is a sense of being *understood* by his or her parents, irrespective of similarity or dissimilarity of their psychological traits. Tooley assumes that such parental understanding – of how things look from a child's point of view – will be secured by psychological similarity, but lacks reason to assume that similarity is necessary or sufficient for parental understanding. After all, the kind of understanding relevant here is not simple comprehension but something more aptly described as sympathetic appreciation. We should also acknowledge that a parent's *perception* of psychological similarity to her child is not necessarily benign. While it is possible that a parent who suspects such a biologically based psychological similarity might be more than usually accepting of the particular recognized trait, that is by no means guaranteed, and moreover, parental expectations for a child to 'follow in their parent's footsteps' may increase as a result of perceived psychological similarity, again potentially threatening a child's burgeoning autonomy and, perhaps, their right to an open future.[7]

In their readiness to assume a strong relationship between psychology and genetics, and to ascribe considerable moral value to biological relationships, psychic similarity arguments embrace an exaggerated view of the significance and value of genetics and biology. This valorization of genetic factors can be regarded as a form of bias, to the extent that it involves an arbitrary attribution of higher value to certain factors and causes (namely genetic ones) in disregard of others (namely socio-environmental ones). We might label this bias *geneticism*, using the term to denote *the morally arbitrary valorization of genetic factors in the understanding of human nature, selfhood and relationships.* More specifically, geneticism involves the assignment of greatest moral significance to conditions, causes and relationships that are biological and genetic in kind, whilst overlooking, downplaying or minimizing the significance of non-genetic, socio-environmental conditions, causes and explanations.

Importantly, geneticism is conceptually distinct from genetic determinism, the latter being, broadly speaking, the socio-biological view that genes rigidly determine human development and behaviour. As suggested earlier this view is certainly false, and there are few theorists nowadays who would endorse it in at least its original formulation. However, in spite of widespread rejection of crude genetic determinism we can nevertheless discern, in the kinds of arguments considered here, a residual tendency to imbue biological and/or genetic functions with a significance not strictly warranted or supported. While the question of the exact role played by genes in human development and behaviour is ultimately an empirical one, the attribution of moral significance to biological and genetic factors and relationships is distinctly and inherently evaluative. I have suggested that 'psychic similarity' arguments represent examples of the tendency towards evaluative bias of the geneticist kind. I want now to suggest that the same can be said of a second argument offered in support of biological relatedness, namely that proposed by David Velleman in his recent paper, 'Family History'.[8]

Like the psychic similarity arguments, Velleman's defence of biological relationship is based on a consequentialist estimation of the benefit of such relatedness for children and parents. Velleman's arguments are explicitly directed at anonymous gamete donation, but his general account of the value of biological connection readily lends itself to support for social cloning, since the value that Velleman attributes to ordinary biological relatedness between parents and children would only be compounded where the relationship is as close as it would be in the case of cloning; as close, that is, as that between identical twins.[9] A Velleman-style defence of social cloning would be one that embraces the kinds of assumptions and evaluative bias described in relation to psychic similarity arguments as 'geneticist'.

Velleman defends the view that biological connection is a 'basic good', one 'on which most people rely in their pursuit of self-knowledge and identity formation'.[10] For Velleman, being raised by one's biological parents facilitates self-knowledge and informs one's sense of identity. He claims, '. . . in coming to know and define themselves, most people rely on their acquaintance with people who are like them by virtue of being their biological relatives'.[11] To explain the link between biological relationship and self-knowledge Velleman invokes the notion of a 'self-concept of a family-resemblance kind', possessed by those who have been raised by their biological families.[12] Family resemblance is defined as 'a similarity that can be immediately recognized but not readily analysed or defined'; and Velleman claims that 'much of what [we] know about [ourselves] is contained in this family-resemblance concept and cannot be articulated'.[13] And he also claims that this '. . . family-resemblance knowledge about myself includes information not only about how I look but also about my personal manner, my styles of thinking and feeling, my temperament . . .', '. . . much of my psychological knowledge about myself . . .'; and '. . . much of that self-knowledge by which I am guided in my efforts to cultivate and shape myself.'[14] Possession of a family-resemblance concept is therefore, for Velleman, linked in important ways to projects of self-cultivation and identity formation, as well as to self-knowledge.

Velleman needs to say more, of course, to achieve his desired link between possession of a family-resemblance concept and a *literal* family to whom one bears a literal family resemblance. It is worth quoting him in full on this point:

> I think that forming a useful family-resemblance concept of myself would be very difficult were I not acquainted with people to whom I bear a literal family resemblance. Knowing what I am like would be that much harder if I didn't know other people like me. And if people bear me a literal family resemblance, then the respects in which they are like me will be especially important to my knowledge of what I am like, since they resemble me in respects that are deeply ingrained and resistant to change.[15]

Whatever we might say about the capacity to form a family-resemblance concept of oneself in the absence of knowing people to whom one bears a literal family resemblance, cloning would seem to offer precisely the kind of literal family resemblance that Velleman deems central for the development of a family-resemblance concept, and therefore for self-knowledge. Thus if Velleman is correct, knowing the person from whom I was cloned should provide me with the greatest assistance in my quest for self-knowledge, since my literal family resemblance to the person from whom I was cloned, and to whom I am therefore genetically *identical*, would be substantially greater than my literal family resemblance to the people to whom I am less closely genetically related.

Should we suppose, then, that knowing our cloned parent(s) would provide us with knowledge about ourselves, and in particular our psychological traits and point of view? I think such a supposition would be deeply mistaken. Certainly, being acquainted with those who have parented us, and possessing some understanding of the way in which our parents' traits and viewpoints have influenced the development of our own, can potentially contribute to self-knowledge in fairly significant ways. But, contra Velleman, to acknowledge this is not to identify any necessary or distinct contribution made by the *genetic* connection between our parents and ourselves. I find it hard to imagine that the kinds of things we know about ourselves purely in virtue of knowing about our *genetic* or biological similarity to our parent(s) are the kinds of things that yield self-knowledge of any significant kind. This is because on any meaningful conception of what makes us the selves we are in terms of psychological traits and viewpoints, who we are is not given by our genotype or, more broadly, biology. To think otherwise is to give excessive weight to the significance of genetic factors, and therefore, to fall foul of the evaluative bias of geneticism.

More worrying than Velleman's claims about biological relationships as a source of knowledge about who we actually *are*, however, is his view of biological relationships as a source of knowledge about who we may *become*. He says:

Acquaintance with a child's biological family can be a source of knowledge for people other than the child itself . . . Information relevant to self-cultivation is also relevant to the rearing of children . . . So much of what perplexes parents has to do with the nature whose unfolding they are trying to foster. How far can the child hope to reach, and in which directions? What is the child unable to help being, and what can it be helped to become? What will smooth its rough edges, and what will just rub against the grain?[16]

Extending Velleman's comments to the biological similarity established by cloning, we should ask: could knowing our genetically identical parent provide knowledge of 'what we can make of ourselves', as Velleman puts it?[17] To assume that it would is surely to overstate the role of biological and genetic causes in determining what we can become; it is, again, a geneticist view of human nature and human capacity. Such a view threatens our own ability, and the ability of others, to think of our selves as in important respects constituted *by ourselves*, as *made* in the process of living, not *given* by the facts of our genetic make-up, and not prefigured by what those with genetically identical makeup were able to 'make of' themselves. Any constraints on our potential that are imposed by our families are not constraints imposed by what we share with them genetically. They are much more plausibly thought of as constraints imposed by shared social conditions, expectations and experiences.

In short, one needs to hold a fairly 'thick' biological or genetic conception of individual human nature in order to find something compelling in Velleman's argument of the value of biological relationship. Such a conception is open to challenge by accounts of human nature and self-knowledge that are more compelling for their greater emphasis on the socio-environmental conditions that contribute to psychology, identity formation, self-conception and self-knowledge. Couched in terms of a Velleman-style self-knowledge argument, biological relatedness should not be uncritically accepted as a moral warrant for social cloning.

A concern might arise at this point that to the extent that the above assessment challenges the morality of social cloning, it poses an equal challenge to any means for achieving biologically related children, including, most obviously, IVF. If we are sympathetic to the above grounds for denying moral warrant for social cloning, are we not thereby committed to a view that IVF similarly lacks moral warrant, falling equally foul of geneticism? In response I want to suggest that there is a relevant difference between IVF and social cloning that significantly mitigates concern about the geneticist bias in the former case while leaving it intact in the latter. The difference pertains to the nature of the biological aspirations of prospective parents who seek to use these distinct forms of reproductive intervention. In the case of social cloning, prospective parents are, firstly, seeking to determine the entire genetic repertoire or make-up of their child; and secondly, seeking a child with whom they have a genetic relationship of (near) identity. In the case of IVF, in contrast, parents bring no such exacting ambitions. Although they seek a biologically related child, the closest biological relationship sought is that of mere genetic relatedness, not identity. And given the use of donor material in standard IVF treatment there could certainly be no reasonable parental expectation concerning the precise particulars of any biological or genetic repertoire. In IVF, as in non-assisted reproduction, much is left to chance and the so-called 'natural lottery' of fertilization and embryo development. For these reasons, while genetic relatedness (to at least one parent) is certainly secured in both IVF and social cloning, the psychic similarity and self-knowledge arguments that have been our particular focus in this discussion do not play a role in defences of IVF in the way that they do in defences of social cloning.

Essentialist assumptions in social sex selection: the 'family balance' argument

I have argued that at least two arguments for the moral significance of parent–child biological connectedness lack the foundation they are assumed to have. Indeed, any benefits that might attach to genetic relatedness are benefits only in virtue of widespread perceptions concerning its moral significance and value. I want now to consider an equally problematic evaluative bias in relation to a second kind of sorting practice: namely non-medically based preconception sex selection by means of pre-implantation genetic diagnosis (PGD) and IVF (or 'social sex selection').

Social sex selection has been defended as the legitimate expression of procreative autonomy, described by Julian Savulescu as 'the liberty to decide when and how to have children according to what parents judge is best'.[18] As Dorothy Wertz and John Fletcher point out, extending reproductive control to control over the sex of our children can be viewed as the logical extension of other forms of reproductive control and freedom.[19] According to the most extreme formulations of the procreative liberty argument, it is permissible for prospective parents to use PGD and embryo selection to select for any offspring traits they desire. Like Savulescu, John Robertson argues that procreative liberty should be presumptive, justifiably overridden only where tangible harm to identifiable individuals would result.[20]

Objections to social sex selection are typically consequentialist, focusing on potential harms such as those associated with predicted demographic sex ratio imbalances, and the further discriminatory effects of sex selection practice within societies with strong sex preferences, such as India, China, the Middle East and East Asia.[21] There is sound evidence that sex selection practice in countries with strong prevailing sex preferences has led to considerable harm, including widespread abandonment, abortion and infanticide of females. However, consequentialist concerns about such harms and about possible sex ratio imbalances are generally not considered relevant in relation to the practice of social sex selection in places like Australia, Western Europe and North America, where there is not a prevailing preference for one sex over another.[22] The familiar feminist charge that sex selection is inherently sexist is therefore thought to be disarmed, and on those grounds it is sometimes argued that a prohibition of social sex selection in countries such as these would be unjustified.

I will not here comment on that particular debate, but want instead to give closer consideration than has been customary to a particular defence of social sex selection, one that has been thought to escape the force of feminist objections by avoiding the problematic link between social sex selection, sex preference and sexism. If defences of social sex selection can be construed so as not to embody specific sex preference, they appear capable of avoiding the objection that social sex selection is necessarily sexist.

With precisely that purpose the argument has been advanced that social sex selection is morally innocuous where it is supported by what are commonly referred to as 'family balancing' reasons.[23] Social sex selection aimed at family balancing has been defended on a variety of grounds. Firstly, its potential to inhibit population growth by enabling those who desire a balanced family to achieve it without making more attempts – and consequently more children – than they would otherwise prefer, is widely touted. Secondly, avoiding an *intra-familial* skewed sex ratio is sometimes offered in defence of the preference for a sex-balanced family. It has been claimed that '. . . the presence of siblings of the other sex might promote mutual understanding among the children';[24] and that parents might '. . . want their children to respect sex-based differences and to learn fairness to the opposite sex by practicing it at home.'[25] Others have pointed to an alleged greater parental satisfaction level associated with parenting children of both sexes. Pennings puts the point this way:

[T]he pleasure and happiness generated by the variety of having both boys and girls can be considerable. Being a parent of children of different sexes brings different and more diverse experiences.[26]

Accordingly John Robertson has claimed that it '. . . would be non-sexist to use preconception gender selection [as he terms it] to produce a girl because of a parental recognition that the experience of having and rearing a girl will be different than having a boy.'[27]

Before we consider this view more closely we should acknowledge that although the preference for a family that is balanced in terms of sex ratio is an inherently *sex*-based preference, it does not embody any *specific-sex* preference. This is in spite of the fact that it will involve selective decisions in favour of one particular sex or the other. The important point is that which sex is selected for is, in the context of family balance arguments, an entirely contingent matter and determined by reference to the sex of other children in the family. Of course, for the family balance preference to have this prima facie sex-neutral and therefore supposedly morally innocuous content, there should exist no sex preference in regard to the *first*-born child. Unfortunately Pennings notes that research on birth order preference reveals 'an overwhelming preference for a boy as the first child and a girl as the second child.'[28] While possibly compatible with family balance as the primary motivation for sex selection, the existence of a strong birth order sex preference is, I believe, suggestive of sex-stereotyping, at least, and is vulnerable to the charge that it embodies inherent sex essentialism, about which I will say more presently.

Absent such distortions of the family balance argument, some feminist philosophers have been willing to accept that the preference for a balanced family is not inherently sexist. Mary Anne Warren, for example, presents three positive 'quality of life' considerations that, in her view, provide sufficient support to justify social sex selection for family balance reasons.[29] Firstly, being of the 'wanted' as opposed to the 'unwanted' sex will enhance the quality of life of a *child*; secondly, having the desired balance will enhance *family* quality of life; and thirdly, having to undergo fewer pregnancies and births (to which we should add all of the other labour associated with rearing children) to achieve the desired number of children of each sex will enhance *women's* quality of life.

It is certainly true that *within a context* of sanctioned preferences for sex-balanced families, quality of life considerations are likely to be compelling. The benefits suggested by Warren will indeed flow from allowing people to select the sex of their children within such contexts. However, it is important to see that the merit of these as reasons to justify social sex selection, depends on the very issue that is in question in this discussion: that of whether the family balance preference is itself morally unproblematic. My interest is precisely in asking whether we should accept the balanced family preference at face value, as a consideration of weight in determining the permissibility of social sex selection. The preference for a balanced family is indeed widespread, as Pennings observes, but as noted earlier, the fact that something is valued does not establish that it has value, and should not count as a moral consideration in its favour.

Against assumptions of the morally innocuous character of the family balance preference, I want to suggest that while its underlying basis may not be sexist, it nevertheless falls foul of an assumption, or set of assumptions, that is morally problematic in much the same way as were the assumptions underlying the justifications of social cloning, which I criticized on grounds of their geneticism. In particular, the family balance preference falls foul of assumptions that can be characterized as *essentialist*. More specifically, the family balance argument is essentialist in so far as it assumes that significant sex differences exist, and exaggerates the importance of sex in determining gender, personality and family relationships.

In characterizing these problematic assumptions as 'essentialist' I do not have in mind the biological reductionist view that who or what we are – our identity or essence – is reducible to our sex. Those who regard a sex-balanced family as ideal are not necessarily committed to biological reductionism (just as geneticism is not coextensive with genetic reductionism or determinism). Rather, by *essentialism* I am referring to the *morally arbitrary valorization of sex identity in our understanding of human nature, selfhood and relationships*. It is a weakness in family balancing arguments that they seem to implicitly assume that differences in sex equate to differences in gender, personality, behaviour, relationships and/or childrearing experience. Where social sex selection is sought in order to achieve a girl child to balance a family comprising two boys, for example, the family balance argument appears to assume that to bring a girl into the family is to bring something distinct to the family dynamic, where what is distinct is given *in virtue of the child being a girl*. That distinct contribution is assumed, *in advance*, to counterbalance the distinct contribution made by the boys *in virtue of them being boys*. It is this aspect of the family balance argument that I believe makes it susceptible to the charge of essentialism.

Referring to psychological research on sex difference that offers prima facie support to this kind of idea, Robertson says:

> It has long been established that there are differences in boys and girls in a variety of domains, such as (but not limited to) aggression, activity, toy preference, psychopathology, and spatial ability.[30]

Robertson himself acknowledges that such differences could be inborn or learnt, but claims that either way, '... they are facts that might rationally lead people to prefer rearing a child of one gender rather than another ...'[31] I agree. It might be rational to have such preferences within social contexts that attribute significant difference on grounds of sex, whether accurately or not. However, as I have indicated, our interest ought to be in the question of whether such attributions are themselves well founded, and not simply in whether they are made.

It would certainly be obtuse to deny the considerable impact on gender identity development of social perceptions of the significance of sex difference. Gender roles and stereotypes are prevalent in virtually all societies to this day, and undoubtedly influence children's preferences and expectations as surely as they do the preferences and expectations of those around them. Hence Robertson is correct to point out the importance of recognizing that biological sex is invested with social and psychological meanings for prospective parents and for society more generally.[32] My point is not to deny that, but to ask a deeper question about the extent to which we should accept the widespread social alignment of gender and personality with sex, and the extent to which we should endorse parental preferences and expectations that uncritically accept that alignment. To the extent that parental desires for sex-balanced families are based upon gender role stereotypical conceptions of the differences between boys and girls, social sex selection for family balance reasons may reinforce those stereotypes. Presenting their view that the desire for a balanced family 'assumes sex role stereotyping', Wertz and Fletcher ask, 'Why desire to balance a family unless you already hold stereotypes about sex?'[33]

It may be overly hasty to attribute stereotypical sex conceptions purely on the basis of the presence of a desire for a sex-balanced family. However, Wertz and Fletcher's question at least obliges us to adopt a more critical stance towards the family balance preference, one of willingness to enquire about the extent to which the preference is grounded in assumptions of a significant connection between sex, on the one hand, and gender, behavioural traits, personality, social roles and the like, on the other. Moreover, it obliges us to ask whether the family balance argument does not in fact encourage excessive attention, in our reproductive decision-making, to the importance of sex and sex difference. And it obliges us to consider the extent to which the family balance justification for social sex selection embodies essentialism – however unwittingly – on the part of those who desire balanced families. The existence of significant innate differences between the sexes, sufficient to manifest in the morally significant differences thought relevant for intra-familial relationships and childrearing, is far from scientifically established. Yet our social attributions of difference are robust, and in terms of the implications for gender role stereotyping, feminism has certainly helped us to see that such social attributions are far from benign.

Again, I do not wish here to deny the significance of sex in the formation of the self-conception of a child born in today's world, with its ubiquitous representations of both sex and gender roles. Some have argued that sex differences are properly regarded as no more significant than, for example, differences in eye colour. However, that is not the view endorsed here. As noted above, it would be obtuse to deny the impact of social perceptions of the significance of sex difference in the development of gender identity. That is, however, compatible with the view I am defending here, that it ought not to matter so much which sex one's child is born to.

A distinction might help to clarify the point here, between two possible senses in which we might attribute significance to sex. Firstly, to say that sex has an *ex post* (or 'after the fact') significance would be to say that, in the light of their birth into a society that (rightly or wrongly) places considerable emphasis on sex and sex difference, a child's sex will come to have a contingent significance, both to themselves in their identity formation and to/for others. However to acknowledge that sex has *ex post* significance – the denial of which is implausible – is distinct from according an *ex ante*, or 'before the fact' significance to sex. Family balance arguments do more than acknowledge that once a child is born, whether it is male or female will come to be significant to them, and is therefore understandably significant for its parents. They make more particularized, non-contingent claims about the significance, for particular families, of future children being of a particular sex. As such they approximate something bordering on *necessity* claims about the relationship between sex, gender identity and social roles. The distinct nature of a future child's contribution to the family dynamic is assumed in advance, and it is this aspect of family balance arguments that makes them, in my view, vulnerable to charges of essentialist thinking about sex and sex roles. Undeniably, once a child is born its sex will have meaning and significance. That admission of the significance of sex is, however, very different from a claim that, in advance of a child being born, it is important that it be born of this particular sex.

Conclusion

I have argued for critical apprehension with regard to procreative preferences that – albeit perhaps unwittingly – embrace views of the necessary significance of sex difference. The 'family balancing' argument sometimes offered in support of social sex selection appears to embrace essentialist thinking about sex and gender, and in doing so may reinforce the sex stereotyping that feminists in particular have fought so hard to challenge. Likewise, the 'psychic similarity' and 'self-knowledge' arguments sometimes offered in support of social cloning embody a geneticist bias that I am confident we will in time discover we have equal reason to reject. That social cloning and social sex selection are impermissible reproductive interventions has not been hereby established. But if our support for these practices is to be based on defensible moral grounds, arguments that are free from geneticist and essentialist assumptions will need to bear the justificatory burden.

ACKNOWLEDGEMENTS

Thanks to Janna Thompson and Cynthia Townley for helpful suggestions on earlier versions of this chapter.

NOTES AND REFERENCES

1. This view is defended in N. Levy and M. Lotz, 'Reproductive cloning and a (kind of) genetic fallacy' (2005) 19 (3) *Bioethics* 232–50.
2. In B. Almond, 'Family relationships and reproductive technology' in U. Narayan and J. J. Bartkowiak (eds.), *Having and Raising Children: Unconventional Families, Hard Choices and the Social Good* (University Park, Pennsylvania: The Pennsylvania State University Press, 1999) 103–18.
3. *Ibid.*, 104.
4. M. Tooley, 'The moral status of the cloning of humans' (1999) 18 *Monash Bioethics Review* 27–49.
5. *Ibid.*, 42.
6. Of course, the genetic underpinning of psychological traits remains a matter of some contention, and some are optimistic regarding our prospects for discovering the genetic underpinnings of a vast and indefinite number of psychological traits. See, for example, E. O. Wilson, *Sociobiology: the New Synthesis* (Cambridge, MA: Harvard University Press, 1975) and *On Human Nature* (Cambridge, MA: Harvard University Press, 2004); and S. Pinker, *How the Mind Works* (London: Penguin, 1997).
7. See also J. Feinberg, 'The child's right to an open future' in W. Aiken and H. La Follette (eds.), *Whose Child? Children's Rights, Parental Authority and State Power* (Totowa, New Jersey: Rowman and Littlefield, 1980); C. Mills, 'The child's right to an open future?' (2003) 34 (4) *Journal of Social Philosophy* (4) 499–509; and M. Lotz, 'Feinberg, Mills and the child's right to an open future' (2006) 37 (4) *Journal of Social Philosophy* 537–51.
8. Published in *Philosophical Papers (Special Issue: Meaning in Life)* 34 (2005), 357–78.

9. As Velleman points out, anonymous donor conception involves purposely creating a child who will know only half of its biological parentage, which for Velleman constitutes a disadvantage (*ibid.*, 365). No comparable disadvantage would be imposed on a cloned child, of course.

10. *Ibid.*, 365.

11. *Ibid.*

12. *Ibid.*

13. *Ibid.*

14. *Ibid.*

15. *Ibid.*, 365–6. It is not clear why Velleman suggests that literal family resemblances are 'deeply ingrained and resistant to change', but he may have in mind here properties and traits that are laid down in our biological/genetic make-up and are change-resistant for that reason.

16. *Ibid.*, 370.

17. *Ibid.*, 368, 370.

18. J. Savulescu, 'Sex selection: the case for' (1999) 171 (7) *The Medical Journal of Australia* 373–5. See also R. Rhodes 'Ethical issues in selecting embryos' (2001) 943 *Annals of the New York Academy of Sciences* 360–7.

19. D. C. Wertz and J. C. Fletcher, 'Sex selection through prenatal diagnosis: a feminist critique' in H. Bequaert Holmes and L. M. Purdy (eds.), *Feminist Perspectives in Medical Ethics* (Indiana and Bloomington: Indiana University Press, 1992) 241.

20. J. A. Robertson, 'Preconception gender selection' (2001) 1 (1) *American Journal of Bioethics* 2–9.

21. A recent *Lancet* publication reported that as a result of prenatal sex selection and selective abortion in India alone, there is an estimated 'deficit' of 500 000 Indian female births per year, summing to more than 10 million 'lost' females in 20 years. See P. Jha, R. Kumar, P. Vasa *et al.*, 'Low male-to-female sex ratio of children born in India: national survey of 1.1 million households' (2006) 367(9506) *Lancet* 211–18; and S. S. Sheth, 'Missing female births in India' (2006) 367(9506) *Lancet* 185–6.

22. See, for example, Statham *et al.*'s study of over 2300 pregnant women in Britain, in H. Statham, J. Green, C. Snowdon, M. France-Dawson, 'Choice of baby's sex' (1993) 341 *Lancet* 564–5; and studies with similar results in the USA, such as A. R. Pebly and C. F. Westoff, 'Women's sex preference in the United States' (1982) 19 *Demography* 177–89.

23. See G. Pennings, 'Family balancing as a morally acceptable application of sex selection' (1996) 11(11) *Human Reproduction* 2339–45.

24. *Ibid.*, 2342.

25. D. C. Wertz and J. C. Fletcher, 'Sex selection through prenatal diagnosis: a feminist critique', note 19 above, 244.

26. G. Pennings, 'Family balancing as a morally acceptable application of sex selection', note 23 above, 2342.

27. J. A. Robertson, 'Preconception gender selection', note 20 above, 5.

28. G. Pennings, 'Family balancing as a morally acceptable application of sex selection', note 23 above, 2342.

29. M. A. Warren, 'The ethics of sex preselection' in J. Humber and R. Almeder (eds.), *Biomedical Ethics Reviews* (Clifton, New Jersey: Humana Press, 1985), 730–89. See also her *Gendercide: The Implications of Sex Selection* (Totowa, New Jersey: Rowman and Allenheld, 1985).
30. J. A. Robertson, 'Preconception gender selection', note 20 above, 5.
31. *Ibid.*, 5–6.
32. At Note 1 in G. A. Robertson, 'Preconception gender selection', note 20 above, 8.
33. D. C. Wertz and J. C. Fletcher, 'Sex selection through prenatal diagnosis', note 19 above, 244.

11

The sorting society: a legal perspective

Loane Skene

Many concerns have been expressed about the long-term implications of allowing women to have pre-implantation or prenatal genetic diagnostic tests for their fetus or embryo and the potential medical interventions that may follow from such tests.[1] These interventions include implantation of only 'unaffected' embryos; sex selection; termination of pregnancy if a fetus is found to be affected; pre-implantation or prenatal 'gene therapy'; and even cloning an embryo to avoid mitochondrial genetic disease, which is not avoidable by prenatal testing.

There are concerns about 'playing God', 'designer babies' and judging the 'value' of a baby who is chromosomally or physically abnormal. There are fears about genetic determinism. If genetic testing becomes routine, will undue weight be placed on genetic factors in assessing people's potential achievements, so that those with particular genetic traits suffer unfair discrimination? If it is possible to avoid the birth of children with genetic disorders, will pregnant women feel compelled to agree to tests, or even be penalized if they refuse and have a disabled child, for example, by having to pay higher health insurance premiums? Will the community's sympathy for people with disabilities be reduced as fewer babies are born with congenital disabilities? There are also social justice issues. Who will fund routine genetic tests? If women must pay for the tests themselves, what are the implications for equity? Could genetic tests lead to a society of 'unequals'?

On the other hand, there are many arguments in favour of pre-implantation and prenatal genetic testing. It might be said, as Julian Savulescu has argued in his chapter in this book, that we have a duty to have the 'best' children we can. Parents are free to make other choices to optimize their child's prospects even if that leads to inequity (examples include the provision of private school education). Women will not be compelled to undergo testing. Most disabilities

The Sorting Society: The Ethics of Genetic Screening and Therapy, ed. Loane Skene and Janna Thompson. Published by Cambridge University Press. © Cambridge University Press 2008.

are not congenital but are acquired during life, so there will still be many people with disabilities in the community. Disability will not become so unusual that public sympathy declines for people with disabilities. If fewer children are born needing expensive health care, there will be more resources available for the care of other people.

These arguments and many others have been canvassed in earlier chapters and I have outlined some of them here simply to indicate the range of views about genetic tests on embryos and fetuses.[2] The purpose of this chapter is to consider how the law might be employed if the community decided that genetic tests should be limited in any way. It should be emphasized at the outset that community concern, however deep and widespread, does not of itself require a regulatory response. That is especially the case where opinions are divided. This point was persuasively made by the Australian Committee that recently reviewed the federal legislation on stem cell research and cloning:

> It does not necessarily follow that even though some people think that an activity is unethical, it is necessary to make that activity illegal. Furthermore, the wider the range of ethical views on a particular activity, the weaker the case becomes for declaring that activity to be illegal, with all the attendant consequences of criminal conduct.[3]

However, assuming that the community were to decide that pre-implantation and prenatal genetic testing should be prohibited or limited, how could the law achieve that and what would be the implications of such regulation?

There are two options. One could prohibit or restrict either the initial test or the subsequent termination of pregnancy (TOP). I consider the latter first. It should be noted, however, that preventing TOPs except for very serious conditions will not meet the concerns of those who believe that people should not be entitled to 'choose' the child they will have. The only way to prevent or restrict that choice is to regulate the tests that can be done as women could still have pre-implantation tests and then choose the 'right' embryo, without any need for TOP.

Limiting the grounds on which pregnancy may be lawfully terminated

The law on abortion in Australia varies in each jurisdiction but it generally permits TOP only where there is a real risk to the mother's life, or physical or mental health, if the pregnancy continues. However, that requirement has been liberally interpreted and, at least in early pregnancy, TOP is available virtually on request. Interestingly, in many jurisdictions fetal abnormality is not itself a ground for lawful TOP yet ultrasound examinations are routine and other diagnostic tests are common. The expectation is, of course, that the woman may seek a TOP if the fetus is found to be abnormal and the legal justification is then construed to be a real risk to the woman's life or health from having such a baby.[4]

If there was deep and widespread concern in the community about TOPs being conducted to produce 'designer' babies, it would be possible to amend the legislation in order to prevent TOP on the ground of fetal abnormalities, or of relatively minor fetal abnormalities, or for other specified reasons, either totally, or after a certain stage of gestation. However, it may be difficult to decide the grounds on which TOP might be restricted and at which stage of gestation. This is apparent when one considers the genetic tests for conditions that can be detected only during a pregnancy (i.e. they cannot be detected by pre-implantation genetic diagnosis): see Appendix A. Also, such a prohibition would seem inconsistent with other grounds on which a pregnancy may be lawfully terminated. Despite frequent publicity about Australia's high rate of TOP and the extreme concerns expressed by religious groups in particular, there seems to be little political will to amend the legislation. Indeed, where amendments have been made, the law has become more rather than less permissive concerning the grounds on which TOPs may be lawfully performed. Also, as noted earlier, it would not prevent women 'choosing' their baby by pre-implantation genetic diagnosis followed by selection and implantation of the desired embryo. On the other hand, where conditions can be detected only during pregnancy, the only way to prevent women using genetic tests to 'choose' their baby would be to prevent or limit the prenatal tests that can be conducted, or the grounds on which TOP may be undertaken.

Limiting the conditions for which pre-implantation or prenatal tests may be undertaken

It would certainly be possible to legislate to prevent certain types of genetic tests, both pre-implantation and prenatal tests. The conduct of such tests could be made a criminal offence, punishable by fine or imprisonment. A precedent already exists in Victoria in relation to pre-implantation genetic testing for sex. Under the Infertility Treatment Act 1995 (Vic) it is an offence to undertake sex selection in infertility treatment procedures in order to have a baby of a particular sex except to avoid passing on an undesirable hereditary disorder. Section 50 states:

(1) If a person is carrying out artificial insemination or a treatment procedure, that person must not–
   (a) use a gamete or embryo; or
   (b) perform the procedure in a particular manner–
      with the purpose or a purpose of producing or attempting to produce a child of a particular sex.
      Penalty: 240 penalty units or 2 years imprisonment or both.

(2) Sub-section (1) does not apply if it is necessary for the child to be of a particular sex so as to avoid the risk of transmission of a genetic abnormality or a disease to the child.

A 'treatment procedure' is defined in section 3 to mean artificial insemination or a 'fertilization procedure' which is defined as a medical procedure to transfer an embryo, oocyte, sperm, or oocyte and sperm to the body of a woman. Thus the emphasis is on the *transfer* of an embryo or gametes to the woman's body, rather than *tests* for sex selection but the outcome is obviously the same. What would be the point of undertaking the test if it could not be used in producing a baby of the desired sex?

However, even if there is the community will to create a more general offence in relation to other genetic tests, it would be difficult to draft a provision to create such an offence. The limitation in section 50(1) above – 'with the purpose or a purpose of producing or attempting to produce a child of a particular sex' – is clear and specific.[5] If the section were to be more general, how could the prohibition be phrased? One might possibly say 'except with the purpose of avoiding or minimizing a severe genetic abnormality or disease'. But how would a 'severe genetic abnormality or disease' be interpreted? What conditions would be included? Consider the lists of conditions that can be diagnosed by a prenatal or pre-implantation test: Appendices A and B. Where would one draw the line in prohibiting tests? What difference would it make if the test is pre-implantation (so that there is no issue of TOP); or, if the test can be done only during pregnancy, if it is done early or late in gestation?[6]

If particular conditions are later set out in regulations made under the Act that created the offence, they might quickly be out of date as new genetic tests become available. If a term such as 'severe' is used to provide flexibility as new tests become available, that may be open to question. Many people disagree about the severity of particular conditions. What seems 'severe' to one person is 'within the range of human difference' to another person. If the word 'severe' is omitted from the exemption in the wording of the offence (so that genetic tests are allowed except for 'abnormality or disease'), the prohibition may seem too narrow. One would then need to determine what is a 'genetic abnormality or disease'. This might include even the most minor abnormality, so that the provision would have little effect. It would clearly exclude testing to achieve 'enhancement', rather than 'remediation'; but would it prevent tests to avoid an essentially 'cosmetic' condition, rather than a medical one?

The focus on avoiding 'abnormality or disease' might also be open to question. It would presumably permit tests for late-onset conditions like Huntington disease or familial adenomatous polyposis, even if the risk of the child developing the condition is less than one hundred per cent. But what about tests for carrier or susceptibility status rather than an affected status? Could one say that such a child has an 'abnormality or disease'? And if the offence referred to 'abnormality or disease *in that child*', then that would prevent testing to give birth to a child who is immunologically compatible with a sick child for whom the new child could be a life-saving donor (assuming that such testing is desirable and should be allowed).

Furthermore, a new offence of the Victorian type is limited to testing in the course of infertility treatment. It would not apply to prenatal tests with a view to terminating a pregnancy if the fetus is affected by a particular condition or does not have a desired characteristic. In order to prevent testing of that type, it would be necessary to create an offence of undertaking the test itself. Although I cannot provide a specific precedent for such an offence, it would be possible to draft a legislative provision to create an offence of conducting genetic tests on a fetus (and also on an embryo) other than for certain conditions. The focus would then be on the test, rather than on the transfer of an embryo or TOP after the test. But again, where would one draw the line in deciding which tests should be available and how would one draft the exemptions – and keep them up to date?

The problem might be addressed by agreeing in broad terms about certain conditions for which genetic testing should be permitted, such as those that involve severe conditions, and then prohibiting other tests. Or, having agreed on broad principles to be included in legislation, a committee could be established to determine from time to time the conditions to be included, with further guidance perhaps given by non-legislative guidelines that are easier to change than legislation.[7] That approach would have the benefit of flexibility; the conditions for which testing is prohibited could be changed fairly simply as new tests become available. However, the decision would need to be made after public consultation. It is not desirable to have 'backdoor' changes to legislative policy in such a contentious area, even if the legislation itself envisages that such changes may occur.

Finally, there are problems of jurisdiction. Any legislative scheme is almost always confined to the jurisdiction in which it is enacted. Laws are generally not extra-territorial.[8] This means that women who want to avoid the operation of laws in Australia that prohibit or restrict particular genetic tests could do so relatively easily. If the test is one that the woman can do herself, such as a maternal blood or urine test, then she could take a sample and send it to another country for testing. Alternatively, if the test involves reproductive technology, or it is more invasive for the mother such as amniocentesis or chorionic villus sampling, or it requires testing of a fetus (which will be required for most prenatal tests), the woman would need to go to another country where such tests are allowed. However, there would be nothing to prevent her doing that. This means that genetic tests would again be open to equity objections. The tests would not only be limited to women who can afford the test itself. They would be limited to women who can afford to travel to another country for the test.

Legislating to facilitate, rather than restrict, genetic tests

An alternative use of the law is quite different from the uses that have been discussed earlier. Instead of the law being used to prohibit or limit the genetic tests that may be conducted – or the uses that may be made of test results, or the grounds for lawful TOPs – one might legislate to permit or facilitate genetic testing. It could be argued that the state has an interest in genetic testing that justifies regulation of this kind. The interest might be the protection of the health of women and children; ensuring that high quality services are provided; avoiding costs arising from the birth of children who are likely to require more health services and social security support during their lives, passing on the same needs to their children and the like. Those state interests might justify legislation that requires everyone who conducts genetic tests to be licensed and to report periodically on the tests that have been undertaken and the test results.

Provisions concerning the counselling of women seeking tests and their families could be included in the regulatory scheme. Testing by licensed providers, together with counselling as needed, would then become best practice. Amalgamated test results would provide valuable statistical information about the nature and incidence of genetic conditions and perhaps also the responses of the women being tested (if follow-up information was also obtained). If mandatory licensing was considered too heavy-handed, then there could be a voluntary licensing scheme, with public funding for the test being available only if the test is conducted by a licensed professional.

No specific legislation

The final regulatory option is not to have any specific legislation on genetic testing. It could be left to individual health professionals to decide which tests are appropriate in particular cases. That would enable them to exercise their discretion rather than being bound by the same rule in every case. For example, even if sex selection is considered to be undesirable as a general principle, one might take a different view if a couple have several children of the same sex and want to have another child of the opposite sex. Similarly, one might say that it is generally ethically questionable to give birth to a child in order to benefit another person but if the child's sibling is dying from a condition that could be cured by a donation from a compatible donor and the new child is the most likely person to have compatible tissue, then it would be justifiable to test embryos for immuno-compatibility and choose the one most likely to be beneficial. In the absence of specific regulation, there is scope for compassion in individual cases, leaving it to health professionals to apply a general embargo on tests in other cases.

If the community decides not to have specific laws on genetic testing but is still keen to meet public concern about the potential uses of genetic testing, then professional guidelines could be prepared, perhaps after community consultation, by an appropriate professional association or a government agency.[9] The advantage of guidelines is that they are flexible. They can be updated regularly as new information becomes available. If they are published on a readily accessible website, then people will be able to check them easily. Having professional input, they are likely to be accepted by the health professionals who are bound by them. At a later stage, they can be given extra weight by being included in legislation by reference. If that occurs, they will have the full effect and enforceability of legislation yet retain the advantages of flexibility and professional and community involvement in their initial preparation, monitoring and amendment.

Conclusion

Genetic testing is contentious and there is a wide range of opinion on the tests that should be permitted, whether pre-implantation or prenatal, and the responses that should be allowed, especially TOP. If the community is concerned about the possibility that women will seek genetic tests, either prenatal or pre-implantation, and then 'choose' the baby they want by TOP or by selecting a particular embryo for implantation, the best way to prevent that is probably to legislate to prevent the tests themselves. Trying to prohibit TOP on certain grounds – or the transfer of an unaffected embryo – will not cover all of the activities necessary to prevent attempts to create 'designer babies'. However, drafting legislation to prevent genetic tests will be difficult. There will be questions concerning the types of testing that should be permitted for particular purposes. If the tests are listed, they may soon become outdated. If a general description is used, like 'severe abnormality or disease', it will be open to interpretation and uncertain in its application.

A more flexible approach is desirable. This might involve broad guidelines that have been subjected to extensive community consultation but can still be changed relatively easily in the light of new knowledge. Alternatively, the law might be used to facilitate genetic tests while promoting high quality services.

APPENDIX A[10]

CONDITIONS DIAGNOSABLE ONLY DURING PREGNANCY

- *Anencephaly* – incompatible with life
- *Severe brain damage* – profound and irremediable mental impairment

- *Spina bifida* – high lesion – major physical impairment
- *Organs outside body* – major surgery needed
- *Club foot, cleft lip, cleft palate, dislocated hips* – correctable by surgery

APPENDIX B[11]

CONDITIONS DIAGNOSABLE BY PRE-IMPLANTATION GENETIC DIAGNOSIS (PGD)

- *Trisomy 13* – invariable death in first weeks of life
- *Chromosomal abnormalities* that lead to severe intellectual impairment but prolonged survival. Examples include deletion chromosome 5p, deletion chromosome 4p
- *Down syndrome* – varying prognoses, but invariable mild to moderate intellectual disability
- *Cystic fibrosis* – debilitating condition, early death, intellect unaffected
- *Huntington disease* – adult onset neuro-degeneration, no treatment prevents it but may delay onset or slow progression of condition
- *Hereditary breast cancer* – susceptibility status – uncertain probability and preventive treatment exists albeit potentially, including major surgery (mastectomy)
- *Genetic deafness* – potentially treatable by cochlear implant
- *Haemophilia, phenylketonuria (PKU)* – treatment available but throughout life and significant impact on day-to-day life
- *Haemochromatosis* – treatment may be required in adulthood but minimal impact on day-to-day life
- *Genetic carrier status* (autosomal/ X-linked recessive) – healthy child, risk to their offspring
- *Immunological compatibility* – healthy child born to save dying sibling
- *Sex*, no medical reason – family balance – preference
- *Hair, eye colour* – cosmetic attributes

NOTES AND REFERENCES

1. See, for example, J. Glover (*Choosing Children* Oxford: Clarendon Press, 2006).
2. For Australian experience in pre-implantation genetic testing, see D. Cram and A. Pope, 'Preimplantation genetic diagnosis: current and future perspectives' (2007) 15 (1) *Journal of Law and Medicine* 36–44.
3. Legislation Review Committee, *Reports on the Prohibition of Human Cloning Act 2002 and the Research Involving Human Embryos Act 2002*, 162, para 17.1;. www.lockhartreview.com.au/_files/Legislation%20Review%20Reports%20Full% 20Doc-19Dec05.pdf (accessed 21 August 2007).

4. In Australia, this ground of TOP represents only 1–2% of all those performed but remains far more controversial than others: Dr Carol Portmann, Maternal Foetal Medicine, Royal Brisbane and Women's Hospital, email dated 11 Jan 2005, quoted with Dr Portmann's permission (copy on file with the author).

5. Similarly, one might legislate against attempts to use genetic testing for 'negative enhancement' – to produce a child with a disability. See I. Karpin, 'Choosing disability: preimplantation genetic diagnosis and negative enhancement' (2007) 15(1) *Journal of Law and Medicine* 89.

6. Some tests are possible only later in the pregnancy, such as the test for achondroplasia, which cannot be done before 20 weeks' gestation.

7. In Australia, the Australian Health Ethics Committee of the National Health and Medical Research Council has developed guidelines on genetic testing in its publication *Ethical Aspects of Human Genetic Testing*; an information paper (2000). The Victorian Infertility Treatment Authority also provides general guidelines on its web site; see also D. Cram and A. Pope, 'Preimplantation genetic diagnosis: current and future perspectives', note 2 above.

8. There are ethical issues in relation to extraterritorial laws as well as legal ones; see L. Skene, 'Undertaking research in other countries: national ethico-legal barometers and international ethical consensus statements' (2007) 4(2) *Public Library of Science (Med)*: <http://medicine.plosjournals.org/perlserv/?request=get-document&doi=10.1371/journal.pmed.0040010> (accessed 20 August 2007).

9. For examples, guidelines such as those in note 7 above.

10. I gratefully acknowledge the assistance I received in compiling this list from my colleague Associate Professor Martin Delatyki, Director of the Bruce Lefroy Centre for Genetic Health Research at the Murdoch Children's Research Institute, Melbourne.

   In Australia, newborn screening includes tests for about 30 conditions: Centre for Genetics Education, 'Newborn screening for genetic conditions' The Australasian Genetics Resource Book (2007), Fact Sheet 20 available at www.genetics.com.au/pdf/factsheets/fs20.pdf (accessed 25 February 2008).

11. *Ibid.*

Index